HANDBOOK OF COMMON
METHODS IN LIMNOLOGY

HANDBOOK OF COMMON METHODS IN LIMNOLOGY

OWEN T. LIND

Institute of Environmental Studies
and Department of Biology
Baylor University
Waco, Texas

with 36 illustrations

The C. V. Mosby Company

Saint Louis 1974

Printed in the United States of America

Distributed in Great Britain by Henry Kimpton, London

Library of Congress Cataloging in Publication Data

Lind, Owen T   1934-
    Handbook of common methods in limnology.

    Bibliography: p.
    1. Limnology—Technique.   I.   Title.
QH96.57.A1L55   551.4′82′028   74-8422
ISBN   0-8016-3017-7

GW/S/B   9   8   7   6   5   4   3   2   1

CONTENTS

3 The plankton, 81

INTRODUCTION

The study of limnology is, in its broadest sense, aquatic ecology. The biological limnologist is primarily concerned with the interactions between aquatic organisms and their physical and chemical environments. Because of the more sharply defined environmental limits and further advanced methods of analysis, the limnologist is often able to quantify environmental factors more precisely than is his terrestrial counterpart.

Biologists are occasionally reluctant to attempt essential physical and chemical water analyses because of lack of confidence or even boredom with nonliving systems. This reluctance produces a superficial treatment of ecosystem components, or worse, erroneous physical and chemical data. However, the critical interdependencies of aquatic life and the physical and chemical environment, along with the investigator's awareness of the dynamics of such a system, make total ecosystem analysis not only essential, but fascinating as well.

This handbook is designed for those who wish a collection of basic established methods in a form convenient for use in either the field or laboratory. No chemical procedures requiring more than simple instruments such as the Spectronic 20 or pH meter are included. The appendices, which provide commonly used conversions and equivalents and sources of equipment

1

and supplies, will be especially valuable to the new student or occasional investigator.

Because of rapidly changing instrumentation technology and limnological theory, no completely up-to-date methods book is possible. *Limnological Methods* (Welch, 1948), although dated, contains much useful information, especially on morphometry, plankton, and benthos. It is of limited value in other areas, such as chemistry and biotic dynamics. Fortunately, more recent books on chemical analysis are available. *Methods for Collection and Analysis of Water Samples* (Rainwater and Thatcher, 1960) and, more recently, *Methods for Collection and Analysis of Water Samples for Dissolved Minerals and Gases* (Brown, Skougstad, and Fishman, 1970) are methods manuals used by the Chemical Quality of Water Branch, Geological Survey, U. S. Department of the Interior. The first sections of these manuals are devoted to excellent discussions of collection and analytical techniques as well as instrumentation evaluation.

A preferred source of chemical limnological methods is *Standard Methods for the Examination of Water and Wastewater* (American Public Health Association, 1971; hereafter referred to as *Standard Methods*). This is a compilation of methods agreed upon as standard by three major groups involved in water analysis: the American Public Health Association, the American Water Works Association, and the Water Pollution Control Federation. Wherever applicable, analysts are encouraged to use the tested methods of this book and to report findings in the prescribed terms. This has the advantage of placing the data of many investigators on a comparable basis and allows for more valid comparisons across the country. Often the investigator may simply cite *Standard Methods* in his publications and, because of the wide availability of this source, no further description of analytical methods is required.

A Practical Handbook of Sea Water Analysis (Strickland

and Parsons, 1968) is a widely acclaimed reference book for limnologists as well as those dealing with marine ecosystems. *Methods for Chemical Analysis of Water and Wastes* (Environmental Protection Agency, 1971) is a compilation of methods by the Analytical Quality Control Laboratory of the Environmental Protection Agency for use in the Agency's laboratories. It describes methods that are either those given in *Standard Methods* or modifications thereof as required by the Agency. It also includes techniques to permit automated chemical analysis for many water constituents.

The Hach Chemical Co., Ames, Iowa, has been successful in manufacturing prepackaged reagents for chemical methods of analysis that do not assume any chemical understanding on the part of the investigator. Many of their prepackaged chemical methods are taken from *Standard Methods* and work well.

No single method will be adequate for all waters. Because of unusual concentration ranges of the element of interest, or the presence of unusual and unsuspected interfering substances, the beginning investigator should be especially wary of accepting any analytical results until he has confirmed that the method of choice, whether prepackaged chemicals or detailed laboratory techniques, provides him with correct data. Confirmation of results can be obtained by comparing data with that published from the same or similar waters or by cross-referencing against an alternative method.

Standard Methods also contains sections on bacteria, plankton, and benthos; the latter two discussions are essentially the same as in Welch's book (1948). However, since most aquatic biologists tend to develop and prefer their own techniques for collection and analysis of biological specimens, there are few methods that can truly be considered standard. A welcome addition for the biological limnologist was the English translation of *Methods of Hydrobiology (Fresh-water Biology)* by Schwoerbel (1970). This book is especially valuable to the beginning aquatic biologist because of the comprehensive subject-listed literature references. Several of the International Biological Program handbooks will also be of interest to the freshwater biologist: *A Manual on Methods for Measuring Primary Production in Aquatic Environments* (Vollenweider, 1969), *Methods for Chemical Analysis of Fresh Waters* (Golterman and Clymo, 1969), *Methods for Assessment of Fish Production in Fresh Waters* (Ricker, 1968), and *A Manual on Methods for the Assessment of Secondary Productivity in Fresh Waters* (Edmondson and Winberg, 1971).

There are several small, taxonomically oriented works on freshwater biology, such as those by Needham and Needham (1962) and Eddy and Hodson (1961). The most comprehensive work on freshwater biology is *Freshwater Biology* (Edmondson, 1959), an excellent taxonomic source, especially for invertebrates. Another is *Freshwater Invertebrates of the United States* by Pennak (1953). For identification of algae, the books by Smith (1950), Prescott (1970), and Prescott (1962) will be helpful. Patrick and Reimer's work (1966) is the best for information on diatoms. Fassett (1960) identifies aquatic flowering plants, and Eyles and Robertson (1963) and Correll and Correll (1972) identify plants of southern waters.

Obviously, a serious student of limnology with professional ambitions will want to purchase most of these books for his own library; however, for all but the most fortunate, this acquisition must be a gradual process. It is for the beginning student that this handbook has been prepared. It is designed to provide adequate information for the beginner, but is by no means intended to replace any of the works cited above. Most of the methods contained herein are not mine; they represent a collection of methods that have been found to provide relatively accurate results with a minimum of complicated procedures. Most are seasoned with advice and comments derived from several years of experience with them.

PHYSICAL
CHAPTER ONE # LIMNOLOGY

MAPPING
Mapping a pond or small lake by plane table method

The mapping of lakes is usually beyond the capability of the limnologist, and he must rely on maps made by well-equipped survey teams. Often aerial photographs with known scale will suffice. However, small ponds (generally less than 10 acres) may be accurately mapped with a minimum of equipment. Whenever possible, mapping in winter on sound ice is preferable. Welch (1948) includes an extensive section on map methods from simple to complex.

Apparatus

Plane table and tripod
Alidade (a crude alidade may be made by setting pins in opposite ends of wooden ruler for "sights")
Compass
Ruler, graduated in tenths of inches
Map paper
Hard lead pencils
Round-headed map pins
100-foot steel measuring tape
2 steel stakes, approximately 2 feet long

Wooden stakes, approximately 3 feet long (25 to 100 depending on size of pond)

Sledgehammer

Stadia rod or other painted pole at least 6 feet long

Metal-core clothesline rope, of sufficient length to reach across pond, marked in 5-foot increments

Small boat or canoe

Depth sounding line or electronic depth sounder

Procedure

1. Select the longest relatively straight section of lakeshore for establishing a base line.

2. Drive steel stake near shore for one end of base line. Most of the pond should be visible from this point. Repeat for other end of base line at a point at least 100 feet from first stake and approximately the same distance from shore as the first. Make the base line as long as possible (Fig. 1).

3. Drive a wooden stake at water's edge at every major change in shoreline configuration. Maximum distance between any two wooden stakes should probably not exceed 50 feet. Drive wooden stakes just deep enough to stay in place for mapping. Steel stakes on base line are permanent and should be driven so only 4 to 6 inches protrude. Tie a cloth "flag" to top of every fifth stake to facilitate counting.

4. Attach map paper to plane table. Set up tripod directly over one metal stake at end of base line, and plumb table center to the stake. Level table and align one edge with true north-south line. Lock table in place. Draw true north arrow in corner of map; also indicate magnetic north (Fig. 2).

5. Judge shape of pond and length of base line, and determine appropriate scale. Record scale on map near north-south arrow.

6. Determine position on map that represents end of base line where table is set up. Mark this on map by firmly setting map pin in table. The pin must not move.

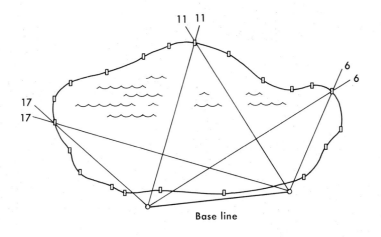

11 11

6
6

17
17

Base line

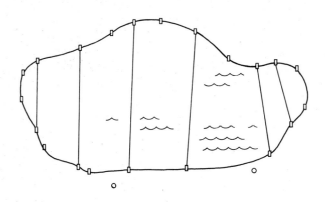

o

o

FIG. 1. Plane table map *(top)* showing position of a numbered shoreline stake and the base line. Distances for three shoreline points as determined by triangulation from the ends of the base line are shown. In a small lake or pond, bottom contours may be determined by sounding along a calibrated rope stretched between approximately opposite shoreline stakes *(bottom)*.

FIG. 2. Student using an open-sight alidade and a plane table to map a small pond. A compass is used to ascertain north orientation of the map. The edge of the alidade at the end nearest the student is always placed against a map pin firmly set in the table. This pin is the map position corresponding to the base line stake over which the table is centered.

Check again to be certain that all parts of pond will fall on map according to scale selected and position of pin.

7. Have flagman set stadia rod vertically on steel rod at other end of base line. Place zero mark on alidade against pin, and sight through alidade down base line to line up with stadia rod (Fig. 2). Draw base line on map along edge of alidade for the appropriate length as determined by scale. *Be careful not to bump the plane table at any time.*

8. Now in a similar manner, progressively sight alidade around the pond on each wooden stake (flagman holds stadia rod just behind stake), and draw line along edge of alidade. Number each line consecutively. The

fifth line and every multiple of 5 should be sighted on a stake with flag. Any deviation indicates a missed stake. After each flagged wooden stake has been sighted and the line drawn, resight on other end of base line. If this sighting does not coincide with original line, the table has been moved and another series must be taken.

9. After a complete circuit has been made, move and reset table over stake at other end of base line. Set pin in map at this end of base line and align table by sighting back down to other end of base line. Repeat step 8.

10. The point where corresponding numerical lines intersect is the position of each wooden stake marking the shoreline. Connect these points with a smooth line. By doing this in the field, you are able to include minor changes in shoreline. Also indicate position of obvious features in the water (logs, weed beds, and others).

11. It is more difficult to map bottom contours for the purpose of morphometric calculations. The problem is the accurate plotting on the map of the position of the sounding team. On smaller ponds, mapping bottom contours may be done as follows: starting at one end of the pond, stretch the calibrated rope between any two known wooden stake locations. Represent this by a light line on the map.

12a. The sounding team in the boat moves along this line, taking soundings at appropriate intervals from one shore to the other (intervals vary depending on amount of change in bottom contours). With the use of the scale, a point is placed at the proper position on the line, and the measured depth written in at that point (Fig. 1).

12b. An alternate procedure, more suitable for larger bodies of water but less precise so far as exact position on the lake is concerned, is the use of timed echo soundings. Echo sounders are becoming increasingly available, especially as "fish finders." The boat's pilot makes a line of sight between two stakes on opposite banks and sets the boat at a constant slow speed. It is important that he make every effort to maintain a straight line and constant speed between the two points. A second person calls out regular time intervals—for example, every 30 seconds—and a third person reads and records the depth shown on the echo sounder at that time. The total time taken to transect the lake is recorded. Thus the total time for distance is known, and time for any timed increment can be calculated. Although this procedure is less precise, it has a compensating advantage in that it allows many more soundings to be taken, which more accurately gives the shape of the lake basin and allows for the plotting of more closely spaced bottom contours. This increases the accuracy of the lake volume calculation.

13. Repeat the chosen sounding procedure for a parallel series of lines across the pond.

14. Determine degree of bottom development to be shown (1-m intervals are often used, but intervals may be more or less). Draw in contour lines by connecting the appropriate points and the lines. For shape between points, parallel the shoreline.

15. Label map and include date and names of map crew. Roll up map (do not fold).

16. Return map to laboratory, retrace shoreline and contours in permanent ink. Clean up other working lines with soft eraser.

Direct measure modification of plane table method

The direct measure modification of the plane table method is suited for ponds of regular outline and open water. It is simple, and depth soundings may be taken simultaneously if the work party is of sufficient size.

Apparatus

Same as for plane table method plus the following
Good quality 18- to 24-inch ruler graduated in tenths of inches
Steel-core clothesline rope, of sufficient length to

reach across pond's longest dimension, marked in 5-foot increments

If approximate size of pond is known before going to the field, a previously prepared table of scale conversions of feet to inches will facilitate time spent in the field

A small boat or canoe may be necessary if obstructions are present

Procedure

1. Select a base point that is relatively open of vegetation and from which all parts of the pond may be seen. Drive iron stake and center drawing board over this stake. All work will be done from this location.

2. Prepare board for drawing as in plane table procedure.

3. Set out shoreline stakes as in plane table procedure.

4. Select point on paper to correspond with plane-table setup over iron stake, and set map pin.

5. Use alidade and take sightings on first stake to left. Draw light line along alidade from base point out to near edge of paper.

6. Place zero end of calibrated line on iron stake, and measure distance to the sighted stake to the nearest foot. Convert this distance to scale in inches; using ruler, measure out from base point the corresponding distance on the line, and mark point. Thus location and distance of first stake have been determined.

7. Repeat step 6 for all remaining stakes.

8. Connect the points, filling in detail from observation.

9. If the map party is of sufficient size, a boat sounding team may work alongside the measuring rope, taking soundings and distance between base point and stake. Depths called out may be marked on map at proper distance from base point (see following section, Morphometry).

MORPHOMETRY

Most limnological phenomena and productivity are directly related to the morphological features of the water basin. Therefore, certain morphometric features are of interest to a limnologist beginning a study of any water. Morphometric measurements are based on good hydrographic maps, and in general, the larger the map, the more reliable the morphometric data that may be obtained from them.

Area by polar planimeter

Whenever instrumentation is available, the polar planimeter method is preferred. The polar planimeter is a delicate instrument, and care must be taken whenever it is used. Read over the directions supplied by the manufacturer for the instrument in use.

FIG. 3. Compensating polar planimeter used to determine areas. Here the planimeter is tracing the shoreline to determine the total area of a pond drawn by the plane table method.

1. Prepare map by placing it on hard, smooth surface. Tape it in place.

2. Most maps will be too large to be covered by one cycle of the planimeter. These maps must be ruled off into halves or smaller segments and the areas of each summed for the total.

3. Check the calibration of your instrument on the map paper being used. Use calibration device supplied by manufacturer, or carefully rule off a known area and trace with planimeter 3 times.

4. Proceed with actual planimetry by tracing outline of lake or bottom contours following carefully the manufacturer's directions.

Area by cut and weight

1. Lightly trace map outline with bottom contours onto a good grade of paper.

2. From an area outside traced area, cut out a square of known area (a 9-inch square is usable) and weigh this piece of paper. Calculate weight per square inch.

3. Cut out outline of entire lake and weigh. Calculate area by dividing weight of entire lake by weight of 1 square inch.

4. Repeat step 3 for each successive bottom contour.

Volume by calculation

If a lake basin is considered as a cone, then the volume may be calculated by the appropriate equation (cone volume = $1.047 \, r^2 h$). However, because the slopes of lake bottoms are rarely regular, a better approximation for volume may be obtained by calculating and then summing the volumes of conical segments (frustra), with upper and lower surfaces delimited by the areas of sequential depth contours. The calculation is then as given by Welch (1948):

$$\text{Lake volume} = \sum \text{frustrum volumes}$$

where

$$\text{frustrum volume} = \frac{h}{3} \left(a_1 + a_2 + \sqrt{a_1 a_2} \right)$$

h = depth of frustrum
a_1 = area of frustrum surface
a_2 = area of frustrum bottom

Volume by planimetry

Lake volume may also be determined by planimetric integration.

1. Using linear graph paper, plot the area at a given depth against that depth. Make the plot with the horizontal axis for area at the top and the vertical axis for depth at the left of the page. This places the 0—0 ordinate in the upper left corner.

2. Integrate the area beneath the curve by using a polar planimeter or by counting squares. Include those squares that are more than half within, and disregard those that are more than half outside the line. If equal di-

Fig. 4. Cartometer in use measuring the shoreline length of a plane table map.

mensions for each square of the grid are used (that is, each grid has dimensions of 1 m^2 in one direction versus 1 m in the other direction), that grid represents 1 m^3 volume. Summing the number of cubic meter volume grids under the curve will give the approximate total volume of the lake.

Shoreline length: cartometer method

The map measure (cartometer) is a convenient method for measuring lengths of shorelines (Fig. 4). It is also a delicate instrument and deserves care in handling.

1. Set dial by turning wheel to zero line. Draw a line of known length on the map paper, and trace three times with the instrument to check its accuracy of calibration.

2. Set instrument to zero line, and carefully trace the shoreline of the lake. Watch carefully to see if the dial revolves more than one time. Record number of inches or centimeters, and convert to feet or meters per scale.

3. Repeat for each of the submerged contours.

Shoreline development

Shoreline development is an index of the regularity of the shoreline. For a lake that is a perfect circle, the shoreline development is 1. As the value departs from unity, irregularity is indicated. This value is calculated as follows:

$$\text{Shoreline development (SLD)} = \frac{S}{2\sqrt{a\pi}}$$

where

S = length of shoreline
a = area of lake

Maximum length and orientation of main axis

These two factors are usually the same portion of the lake, but because of unusual irregularities, they may not be so. Maximum length is the longest straight line that may be drawn without intersecting any mainland.

Not all lakes will have a segment that may be considered to represent a maximum length. The orientation is expressed as opposing points on a 16-point compass: for example, SSE-NNW.

Maximum depth

The maximum depth is the deepest spot in the lake.

Mean depth

The mean depth is an important value, since it is used in certain other calculations, such as heat budgets. It is calculated as follows:

$$\text{Mean depth in meters} = \frac{\text{volume in m}^3}{\text{surface area in m}^2}$$

TURBIDITY

Table 1, a table for turbidity (by Hach Chemical Co.), was made from standard formazin solutions, using a Jackson Candle Turbidimeter to measure the turbidity. The turbidity is expressed in standard Jackson units (JTU).

Apparatus

Spectronic 20 with blue-sensitive photocell (see pp. 37 to 42 for instructions on use)
1-inch matched tubes

Procedure

1. Use demineralized water for blanking the Spectronic 20.
2. Mix sample well by inversion; avoid shaking that would introduce bubbles.
3. Fill a 1-inch test tube with some of the turbid water, and read percent transmittance.
4. Find the turbidity units from the table.

VERTICAL ILLUMINATION

Vertical illumination is defined as the illumination on the horizontal plane at some depth in the water, as

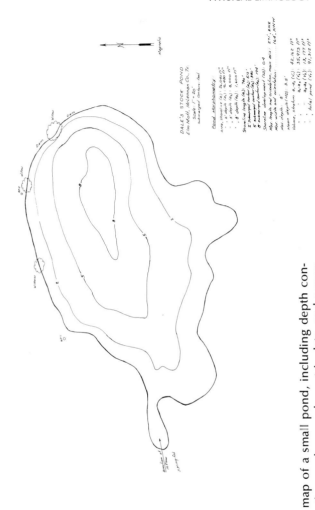

FIG. 5. Completed map of a small pond, including depth contours, north orientation, scale, morphometric data, and names of the mapping party.

TABLE 1. Jackson units (450 nm, 1-inch test tube)

Meter readings	0	1	2	3	4	5	6	7	8	9
10	395	380	360	350	335	320	310	300	290	280
20	273	265	258	250	245	240	233	228	222	217
30	211	206	200	197	193	188	184	180	175	172
40	168	164	160	157	153	150	147	144	140	137
50	134	131	128	125	123	120	117	114	112	109
60	106	104	101	99	97	95	92	90	88	86
70	84	81	80	77	75	73	71	68	65	64
80	61	59	56	54	51	49	47	44	42	39
90	36	32	30	26	22	20	16	12	8	4

measured by submarine photometer. Transmission of light energy through the water column is of special interest for many studies, especially those related to photosynthesis, where one often wishes to determine a depth receiving 1% of the surface illumination. Several factors affect the transmission of light in water. Extinction of light is due to one or more of three factors: the water itself, suspended particles, and dissolved material or color. The intensity of light penetration to depth z may be expressed as:

$$I_z = I_o e^{-(n_w + n_p + n_c)z}$$

where

n_w, n_p, and n_c = extinction due to each of the three factors, respectively

Poole and Atkins (1926) take the vertical extinction coefficient (n'') to be based on penetration of illumination disregarding the angular height of the sun. The validity of this assumption lies in the fact that as the sun becomes lower in the sky and fewer direct rays penetrate the water, a much greater proportion of sky radiation enters the lake with compensating effect. This can be shown empirically to be a valid assumption. This vertical extinction coefficient (n″) is a measure of the slope of the line that is obtained by plotting submarine photometer readings on the logarithmic axis against depth on semilogarithmic paper.

Pure water transmits light maximally at 460 nm in the blue region. Transmission in the red region at 680 nm is almost 70 times less. In natural waters the upper meter serves as an effective filter, removing much of the infrared and ultraviolet radiation. Transmission in the purest lakes approaches that of distilled water, with blue light penetrating to the greatest depths; this, plus the blue of the reflected sky, produces the apparent blue color of these clear lakes. With increasing turbidity, the peak of maximum transmission shifts toward the orange-red portion of the spectrum.

Submarine photometers, equipped with a selenium photovoltaic cell, have a response similar to that of the human eye but extending somewhat further into the blue region of the spectrum and slightly further into the red. The photovoltaic cell has a maximum sensitivity at 555 to 560 nm. It is important to keep in mind that this photocell is not equally efficient at all regions of the spectrum. For example, it is only approximately one fourth as efficient for light with a wavelength of 435 nm as it is at maximum sensitivity.

Some instruments include both a deck cell and a submersible cell. This allows intensity at any depth to be related to surface intensity at the same time. Having both types of cells is especially helpful when surface intensities are variable because of broken and moving cloud cover. Other instruments with only submersible cells adequately relate subsurface intensity to surface intensity by taking surface measurements at the beginning and at the end of the submerged series.

Procedure for single-cell instruments

Follow specific instructions of photometer manufacturer.

1. Open meter and check to see that controls are set at off and at the lowest amplification.

2. Adjust photocell so that it hangs horizontally from the suspending cables. If neutral density or colored filters are used, make sure that all surfaces of glass are wet and free from bubbles.

3. Lower photocell into the water so that its surface is just wet. Record surface reading.

4. Lower photocell into the water to a second pre-scribed depth—usually ½ m will suffice—and take a second reading. Continue this procedure to a depth at which no further light is detected.

5. Return photocell to the surface in the same incre-ments, and retake readings on the way up. Record these values. Any large discrepancy from those obtained while lowering should be rechecked. Most investigators choose to take the average between the two values for the final light reading. Exercise caution when lifting the cell from the water to be sure that it is in a position where the light will not be of sufficient intensity to cause the meter to go off scale; it is usually best to set the function switch to the off position when handling the photocell in the boat.

6. Repeat the entire above procedure, substituting colored filters and removing neutral density filters as required. If possible, use red-, green-, and blue-colored filters.

Plotting extinction curve

In plotting extinction curves one always plots depth on the vertical, or Y, axis. Surface, or zero, depth is taken as the top of the figure, and depths are plotted downward as they naturally occur. With linear paper, a plot of photometer readings against depth results in a smooth curve. An alternate plot producing a nearly straight line results when semilogarithmic graph paper is used. Plot meter readings against depth, placing the meter readings on the logarithmic axis.

Determining vertical extinction coefficient

The vertical extinction coefficient (n''), which describes the slope of the line determined above, may be calculated by the use of the relation:

$$I_z = I_o e^{-n''z}$$

where

I_z = light intensity at depth z
I_o = light intensity at surface
$n'' = n_w + n_p + n_c$

The least square estimate of n'' is given by

$$n'' = \frac{\ln I_o \left(\sum z \right) - \sum \left[z \left(\ln I \right) \right]}{\sum z^2}$$

To solve this equation, divide a sheet of paper into five columns. Column one, to be titled z, is the column in which you will list all of the depths at which light readings were taken, starting with zero as the surface; column two, titled I_z, is the column in which you will list the corresponding intensities at each depth; column three, titled $\ln I_z$, is the column in which you will list the corresponding natural log for each intensity at depth z; column four, to be titled $z \left(\ln I_z \right)$, is the column in which you will list the factor obtained for column three multiplied by column one; the fifth column, to be titled z^2, is the column in which you will list the square of each depth. Columns one, four, and five are to be summed.

The extinction of light between any two depths may be calculated more simply by the following equation:

$$n'' = \frac{\ln I_{z_1} - \ln I_{z_2}}{z_2}$$

This type of calculation allows for the detection of abnormally high or abnormally low light-altering strata within the total column of water when several of these values are compared.

VISIBILITY

Visibility is a measure of the depth to which one may see into the water. Obviously this is variable with the day conditions and the eyesight of the observer. The Secchi disk (Fig. 6) is a simple device used to estimate this depth. It consists of a weighted circular plate, 20 cm in diameter, with the surface painted with opposing black and white quarters. It is attached to a calibrated line by a ring at the center so that when held by the line, it hangs horizontally. To determine the Secchi disk visibility, slowly lower the disk into the water until it disappears, and note this depth. Lower the disk a few more feet, then slowly raise it until it reappears, and note this depth. The average of these two readings is taken for the final Secchi disk visibility depth.

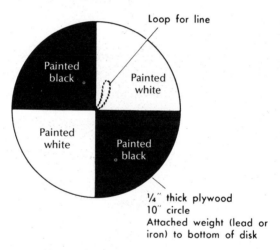

FIG. 6. The Secchi disk may be easily made from a 20-cm diameter metal or weighted wooden disk. Opposite quarters are painted gloss white and gloss black as shown. It is important that the calibrated line be attached so that the disk hangs horizontally in the water. (From Vivian, V. E.: Sourcebook for environmental education, St. Louis, 1973, The C. V. Mosby Co.)

The Secchi disk visibility is useful as a means of comparing the visibility of different waters, especially when measured by the same observer. Since clearness of the day, position of the sun, roughness of the water, and the observer all are significant considerations, they should be recorded along with the visibility depth data. Most important is for an observer to establish a standard set of operating conditions for himself; for example, always take readings while standing, with or without glasses or sunglasses, on the lee side of the boat with the sun to the observer's back, sometime between 9:00 AM and 3:00 PM.

Since one of the more frequently used optical relations in water studies is the photic depth, or depth of 1% surface illumination, some limnologists find it convenient to "calibrate their eye" to estimate photic depth by using only the simple Secchi disk. This is done by accurately determining the true photic depth by use of a submarine photometer and at the same time taking a series of Secchi disk readings to obtain an average. Dividing true photic depth by Secchi disk visibility depth will obtain a factor by which Secchi disk visibility depth is multiplied. This factor is used in the future to estimate photic depth when a submarine photometer is unavailable. If this procedure is used over a range of different water conditions, one general factor may be developed and a fair approximation of photic depth obtained.

TEMPERATURE

Sufficient and accurate temperature data are important to the limnologist. Temperature directly and indirectly exerts many fundamental effects on limnological phenomena such as lake stability and biotic metabolism.

Equipment

Many satisfactory methods for determination of surface and subsurface temperatures are available, the main differences being in convenience of use. Most limnologists prefer to use the electric thermistor thermometer when one is available. However, a weighted, low-cost minimum-maximum thermometer provides an economical substitute. Because of the high specific heat of water, one may even bring samples to the surface and quickly measure temperature with a laboratory thermometer before the water mass has time to change temperature significantly. The brass-cased reversing thermometer was long a standard tool, but because of its relatively high cost and the lowering of prices of thermistor types, it is used much less today.

Procedure: thermistor thermometer

1. With the probe disconnected and the instrument turned on, set the mechanical adjustment of the meter movement, which is usually a screwdriver adjustment located on the case of the meter, so that the meter pointer is aligned with the scale mark. For maximum accuracy on all adjustments and readings, tap the meter *lightly* with the fingernail to overcome any friction in the meter movement, and place the instrument in its normal operating position.

2. Connect the probe and take a shaded air reading of temperature. Then place the weighted end of the cable into the water at the surface, and hold it in place until meter comes to a complete stop. This may require some time on the first reading, since the metal weight tends to hold heat. Lower the cable and take readings at 1-m intervals unless the specific type of data required calls for more frequent or less frequent readings.

3. For maximum accuracy, take a second series of readings while returning the probe to the surface, and take the average of the two measurements as the temperature for that depth.

4. Temperature data are frequently presented as a depth-temperature plot. Since the original definition of a thermocline was any depth where temperature fell

1° C or more for each 1 m of depth, the usual plot uses equal units for temperature (° C) increasing from left to right on the upper abscissa and depth (m) increasing from zero at the upper left on the ordinate. Any slope equal to or greater than 45° from the horizontal may be defined as thermocline. In this context, one should refer to Hutchinson (1957) for a discussion of the metalimnion concept as a better interpretation of a depth-temperature profile.

CHEMICAL

CHAPTER TWO

LIMNOLOGY

COLLECTION OF WATER SAMPLES
FOR CHEMICAL ANALYSIS

Since the mass of water being investigated is usually very great relative to the sample, the sample is representative of the mass only to the extent that the mass is homogeneous. Such homogeneous conditions do not usually exist in the dimensions of either surface area or depth. Recent studies using aerial and satellite photography and imagery have made us especially aware of the nonuniform conditions that exist over the horizontal dimension of the lake surface. Thus, care must be exercised in the selection of the location and number of sampling sites. Frequency of sampling varies with the intent of the investigation. For general information-gathering surveys, bimonthly series may suffice; however, although many investigations are based on even fewer samples, 24 can hardly be considered representative of the environmental conditions in 1 year.

WATER SAMPLERS

Although one may take water samples near the surface by simply filling a bottle or may dive to an approximate depth and open a bottle, a water-sampling device is desirable (Fig. 7). Such devices are designed not only to sample water from discrete depths but also to allow transfer of this water into the storage bottle without agitation or aeration. Several types of sampling bottles have been designed, including the Kemmerer, the Nansen, and the van Dorn. These are available in both metalic and plastic models. They differ principally in closing mechanisms. All are designed to cut a column of water so that when they reach the desired depth, they contain a sample of water from that depth. They are then tripped by a messenger and snap shut in various ways.

A sampler that is adequate for all collections except for dissolved gases may be fabricated out of common materials as follows (Fig. 8). Obtain a glass bottle of 1/2- to 1-gallon capacity. A bottle with a finger loop on the neck is ideal. Affix a weight to the bottom of this bottle sufficient to produce a negative buoyancy when filled with air. This weight can be made by filling a coffee can with concrete. Bore a hole in a rubber stopper that fits the mouth of the bottle. Bolt an eye bolt through the stopper so that the eye is on top. Affix the sampling line to the eye bolt approximately 18 inches from the end, and affix

FIG. 7. Three types of samplers used to collect subsurface water samples. **A,** The Kemmerer sampler has long been a favorite of limnologists because of its ruggedness and reliability. Although usually constructed of brass it is now available in plastic (PVC) material for use where metal ions must be avoided.

Continued.

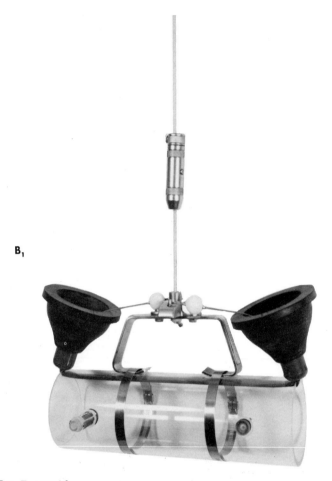

B₁

FIG. 7, cont'd.
B, The van Dorn sampler is usually made of PVC. It has the advantage, because of the external trip mechanism, of being attached to a line in series for winch and boom operation in deep waters. A transparent view **(B₁)** shows rubber tubing holding rubber closures in place. The van Dorn type is available in horizontal models, which permit sampling at discrete depths, especially useful in shallow water or near the lake bottom. A sampler rigged for horizontal use is shown in the "set" position **(B₂).** It is possible to fabricate a van Dorn sampler from PVC pipe, two rubber plunger (plumber's helpers) heads, a length of Latex tubing, and a trip mechanism.

B₂

C

FIG. 7, cont'd.

C, The Nansen bottle, easily attached in series to a line, is an established favorite of oceanographers and large-lake limnologists. Since closure is by rotation with the sampler inverted, Nansen bottles are frequently fitted with reversing thermometers so that exact temperatures at the depth of each sampling are measured. (Courtesy Wildlife Supply Company, Saginaw, Michigan 48602.)

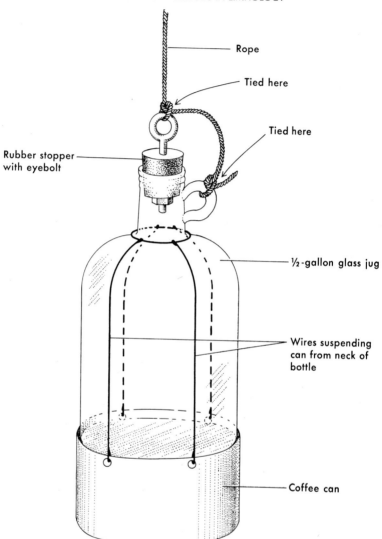

Rope

Tied here

Tied here

Rubber stopper
with eyebolt

½-gallon glass jug

Wires suspending
can from neck of
bottle

Coffee can

FIG. 8. A simple water sampler of the Meyer type is easily constructed from a weighted bottle. This type of sampler is not suitable for collection of subsurface water for dissolved gas analysis.

the end to the neck of the bottle or to the finger ring on the neck.

In use, the bottle is stoppered and lowered gently to the desired depth while suspended from the line attached to the eye bolt in the stopper. When the desired depth is reached, a sharp tug on the line will pop the stopper from the bottle, and the bottle will be suspended from the line secured to its neck. The bottle then fills at that depth.

An alternate procedure that is suitable for gas analyses that do not require extreme accuracy includes the use of the same type of weighted bottle suspended by a line about the neck. In this case a two-holed stopper is inserted in the neck of the bottle, with a small-bore glass tubing projecting through the stopper and reaching almost to the bottom of the bottle. In use, this bottle is lowered swiftly to the desired depth. Water flows in through the small-bore tubing; and if the depth is not too great, little contamination with upper water occurs. The water flowing through the tubing into the bottom of the bottle displaces the air above it, producing a minimum amount of agitation and surface contact with the air. It must be emphasized, however, that this method is crude and does allow for some gas contamination both at the air-water interface in the bottle and through the bubbling of the air out the hole at the mouth.

SAMPLE STORAGE BOTTLES

Unless specifically stated otherwise, samples may be collected and then stored in Pyrex or polyethylene bottles, but soft glass bottles may suffice if analysis is to be performed within 48 hours. New bottles (and other new glassware) should be filled with water and allowed to soak for several days to leach out any water-soluble substances in the glass. Sample bottles must be cleaned after each use as described under the section on glassware.

In the field, the sample bottle should be rinsed with

the water to be sampled before it is filled. Certain factors, such as dissolved gases, are especially susceptible to change after sampling; for these, analysis must be done in the field, and this is so noted with the procedure. Most other factors are subject to varying degrees of change between time of collection and time of analysis. Biological activity within the sample is a major problem. It is recommended that samples be placed immediately in an ice chest, covered with crushed ice, and the lid replaced to exclude light. These may be transferred to a refrigerator in the laboratory while awaiting analysis. However, the temperature specified for the procedure must be noted and time allowed for the sample to warm to this temperature.

LABORATORY PROCEDURES

Good technique must be emphasized. The key to this lies in the skill, experience, and most important, the attitude of the technician. Many data have been discarded after a careful inspection of suspect data has indicated a hurried and careless technique or the lack of attention to detail. One of the best places to develop technical proficiency is in the care of analytical glassware. Many analyses are for trace amounts of an element in very diluted solutions (microchemistry on macroquantities). To prevent contamination, glassware must be spotless. A routine cleaning procedure is as follows.

1. Always empty sample water out as soon as you are sure you are finished with it; rinse bottle with tap water. This prevents the formation of silt and other scums on the bottom, which are difficult to remove.

2. Wash bottle with warm water and a special low-sudsing, phosphate-free glassware detergent. Use brushes when necessary, but avoid scratching with the wire core because any roughening of the surface increases the probability of contamination.

3. Rinse three times with tap water.

4. Rinse with 5% HCl if analysis is for any element likely to be found in detergent.

5. Following acid rinse, rinse three more times with tap water.

6. Give a final rinse, inside and out, with distilled water.

7. Dry and store in a dust-free place.

In the event that the above procedure does not satisfactorily clean the glassware, a chromic acid cleaning solution may be used. This is prepared by adding 100 ml concentrated H_2SO_4 very slowly and carefully to 3.5 ml saturated sodium dichromate solution. *Handle this with caution.*

Before beginning any analytical procedure, always read through the entire procedure. Such a practice will save many samples that would otherwise be ruined and lost. It will also save time, since many procedures call for prior preparations or waiting periods during which time other tasks may be done. Many of the reagents used in these procedures are poisonous and require special precautions. It does you no good to read that you are not to pipette some poisonous substance by mouth after your mouth is full of it. The many concentrated acid and alkaline solutions used in the limnology laboratory are very hard on both the skin and clothing; handle with care, and always *wipe up spills immediately.*

Your determinations are only as good as your reagents; thus take special care to avoid contamination. Keep the glassware covered with the proper stopper for that particular reagent. Do not use glass stoppers for alkaline solutions or rubber stoppers for organic solutions. Do not pipette directly from stock bottles; pour out the approximate amount needed into a clean beaker or flask, and pipette from it. Never return excess contents of such beakers and flasks or burettes and pipettes to the stock bottle. It is false economy to chance the ruin of a liter of reagent in order to save 10 ml.

TERMINOLOGY

Distilled water is water that has been obtained from an ordinary metal still. This water may contain ammonia and carbon dioxide gases, as well as traces of metals. It is used for routine dilutions where the contaminations cited above will not be a problem and for the final rinse of glassware.

Glass-distilled water is distilled water that has been run through a Pyrex glass still. This process effectively removes most trace metals, but ammonia and carbon dioxide persist.

Demineralized water results from the passage of distilled water through a mixed-bed ion exchange column. This produces high-quality water that is essentially free of ammonia but still contains carbon dioxide. This water is called for in the preparation of all reagents unless otherwise specified.

Carbon dioxide–free water is obtained by vigorously boiling demineralized water for a few minutes and cooling without agitation just before use.

Volumetric glassware may be calibrated either "to contain" (TC) or "to deliver" (TD). TD glassware will deliver only when the surface is absolutely clean so that no drops cling to the surface and when the tips of pipettes and burettes are absolutely smooth and free from chips. The contents of TD pipettes with a complete encircling frosted glass ring around the upper portion must be blown out for accurate delivery. The contents of all other TD pipettes (the more common) are allowed to drain by gravity until a small amount remains in the tip; this tip is then touched to the side of the receiving vessel, and a small amount more flows out by surface attraction. This type of pipette always retains a small amount in the tip that is not to be blown out. Volumetric pipettes and flasks are used when the procedure calls for a specified volume in milliliters, such as 1,000 ml, 250 ml, or 1.00 ml. Though not as accurate as volumetric pipettes, accurately calibrated serological pipettes may be

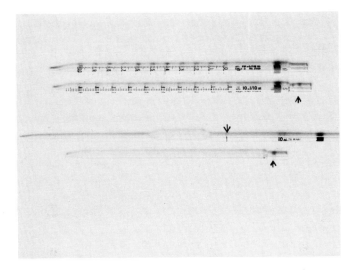

Fig. 9. Commonly used pipettes. The upper two are the serological type, and the lower two are the volumetric type. Both serological pipettes are calibrated to deliver (TD); the second is calibrated to the tip. To deliver 10 ml from this pipette, one must blow out the contents. This requirement is indicated by the encircling frosted glass ring (arrow). The upper of the two volumetric pipettes is the most accurate of the four pipettes for delivering 10 ml. The calibration line (arrow) is in the small-diameter neck, which contains only a small percent of the total volume; thus any error in reading the meniscus is minimized. The contents of this pipette are not blown out, but drain when the pipette is in the vertical position with the tip against the side of the receiving vessel. A small amount of liquid must remain in the tip. The lower volumetric pipette is used to transfer volumes where rapid dispensing is required. It has a large-bore tip, and the contents must be blown out (note frosted ring).

used for measuring fractions of milliliters. When the volume is specified only as 1 liter or 10 ml, then other types of measurements may be used, such as graduated cylinders, where a tolerance of ±5% is allowed.

In order that data may be uniformly interpreted by any investigator it is necessary to use some standard as to how many decimals the results of analysis may be carried; in other words, the investigator must report only *significant figures*. The accepted procedure is to report the certain figures and only the first uncertain figure. For example, if a result of 11.17 ml/liter is read from a curve but it is not known that the procedure is this accurate, the result should be reported as 11.1. If the investigator was certain of the 11 but was uncertain of the .1, the .1 could well be .0 or .2. If, on the other hand, experience has shown the procedure to be accurate and precise to an extent that 11.17 ml should have been reported, then the figure must not be rounded off to a small number as some may do in order that it fit into a column of figures. *The report should include figures justified by the accuracy of the work*. When figures are listed in a column, unwarranted zeros so that all will have the same number of places to the right of the decimal should not be included. However, zeros that are justified should not be dropped; if a burette is accurately read to 9.50 ml, the figure should not be rounded off to 9.5 ml. *Standard Methods* contains more discussion of this important subject.

Precision is a measure of the reproducibility of a method on a given sample under the same conditions. It is usually expressed as the standard deviation from the mean.

Accuracy is a statement of the error of the method and is usually expressed as a comparison of the amount of an element detected by analysis with the amount actually present. This is determined by analyzing for known standard quantities.

INTRODUCTION TO SPECTROPHOTOMETRY

One of the most important laboratory instruments in water chemical analysis is the spectrophotometer-colorimeter. The performance of the Spectronic 20 is between that of the simple colorimeter and the fine spectrophotometer. The ordinary colorimeter uses filters and therefore light from a large portion of the visible spectrum; a spectrophotometer separates light more

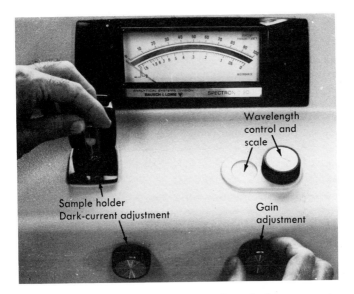

FIG. 10. The easily used Spectronic 20 colorimeter-spectrophotometer. Note the mirrored meter scale. The meter is correctly read when no reflected meter needle image is seen. The small sample holder *(left)* may be replaced with holders accommodating sample tubes up to 1 inch in diameter, thus increasing sensitivity in quantitative analysis. Instrument controls are off-on and zero (dark-current) control *(lower left)*, the 100% transmittance (light or gain) control *(lower right)*, and the wavelength control *(right center)*.

exactly by a prism or grating. The best spectrophotometer uses nearly monochromatic light with a wave band width of less than one nanometer. The Spectronic 20 is a spectrophotometer whose band width is a constant 20 nm over the entire visible spectrum. While this is far from monochromatic, it is adequate for many analyses and has become the workhorse of most laboratories.

The concentration of many substances may be determined by this instrument when a color is produced proportional to the concentration of the substance. The basis for this assumption rests in the Beer-Lambert Law. The absorption of monochromatic light by a solute may be represented by the expression

$$I_z = I_o e^{bz}$$

where

    I_z = intensity of transmitted light
    I_o = intensity of incident light
    e = base of the natural logarithms
    b = absorption coefficient due to the molecular extinction of light and the molar concentration of the solute
    z = thickness of the absorption cell in centimeters

In practical terms, this means that light passing through a given solute is absorbed in a constant exponential manner and that the total amount absorbed is a function of molar concentration, since other factors are constant for that solution. This relationship holds true in its most exact sense only for monochromatic light. When the Spectronic 20 is used, some deviation from a straight relationship must be expected. This difficulty is overcome by establishing a standard reference curve for each analytical method. At regular intervals, or when a new reagent or set of reagents is made up, or whenever the results are suspect, a complete set of standards and a new calibration standard curve are prepared. The new standard curve consists of a *minimum* of 5 standards covering the range of concentrations expected in the water sample. It is wise to prepare one standard near the upper end of the curve each time an analysis is performed to check for deviation from the standard curve.

Errors in the use of the Spectronic 20 that produce invalid results often arise from one of the following.

1. Unmatched test tubes or careless orientation of the tube in the holder.

2. Fluctuation in the light source due to changes in line voltage. This may be partially overcome by the use of an external voltage regulator. However, the analyst may find it best to avoid using the instrument during times of probable variation of demand on the electric lines (4:00 to 5:00 PM is often a poor time). The problem of fluctuation is apparent in failure of the instrument to hold a constant setting of zero absorbance.

3. Fatigue of the photocell. This is also evidenced by difficulty in maintaining a constant zero setting.

4. Error in reading the meter. It must be emphasized that the meter is not uniformly accurate over the entire scale. At high absorbancies the scale is crowded in terms of concentration, so that a considerable change in the relative concentration of the substance will cause only a slight change in the position of the needle. At very low absorbancies, slight differences, such as those produced by fingerprints, dust, bubbles, or faulty positioning, will cause a great change in the meter reading. It is preferable to use the center of the scale, between 1 to 0.1 absorbance, by diluting or concentrating the sample, if possible, or by changing the length of the light path.

5. Color or turbidity present in the sample. In many natural waters, turbidity is a major problem. This includes both natural clay turbidity and chemically produced turbidity resulting from the analytical processes that may cause some other substances in the sample to precipitate. Natural turbidity is best controlled by filtration where possible. Photometric compensation may also be used to correct for interference due to color or turbidity or both in *some* samples. This procedure involves the use

of a duplicate water sample as a blank in place of the distilled water–reagent blank. This sample blank is carried through all steps of the procedure except the one that causes the development of the measured color. Consequently, natural turbidity or color, or artificially produced turbidity caused by chemical changes up to this point, is compensated for by using this sample blank to zero the instrument. Note that this compensation procedure is not universally applicable and that its use must be demonstrated for each analysis. The interferences brought into the analysis by "natural" water causes, such as turbidity or humic acid colors, generally are a greater problem in the blue end of the spectrum due to their higher absorbancies there. It is preferable to work with a wavelength above 500 nm whenever possible.

In an instrument that does not use the entire visible light spectrum, it is necessary to select the best wavelength of light for the particular analysis. In general, this is the wavelength that has the greatest spread of readings between a standard and the blank; that is, the wavelength at which greatest absorbance occurs. However, possible interference from foreign substances in the sample must also be considered, and it may be necessary to select an alternate wavelength free from interferences even though the absorbance may not be as great. For example, in acetone extract of leaf material, chlorophyll a absorbs strongly at both 665 nm and 435 nm; however, carotenoid pigments absorb strongly at 480 nm and interfere at 435 nm for chlorophyll. Thus chlorophyll a is usually determined at 665 nm. For many analytical procedures, the best wavelength may be found in the literature.

For wavelengths above 625 nm, it is necessary to use a red-sensitive phototube and to insert a red filter. Both of these must be absolutely free from dirt and fingerprints when installed in the instrument.

Calibration of matched tubes

Matched sets of tubes may be purchased, or sets may be made in the laboratory using borosilicate glass test tubes in the following procedure.

Apparatus

Spectronic 20 with blue phototube in place set at 510 nm
25 one-inch test tubes
25 one-cm test tubes
Wood or plastic test tube rack
Cloth towels
Glassware for reagents

Reagents

1% HCl. Add 1 ml concentrated HCl to about 40 ml distilled water in a 100-ml volumetric flask, mix, and make up to mark.
Cobalt chloride solution. Dissolve 2.2 g cobalt chloride in 1% HCl in 100-ml volumetric flask; make up to mark with 1% HCl.

Procedure

1. Wash and dry all test tubes carefully. After washing, handle only by tops and place in rack or wrap in cloth towel so that glass does not touch glass.

2. Starting with the 1-cm tubes, add about 5 ml of 1% HCl to one tube to serve as a blank.

3. Add similar volumes of cobalt chloride solution to each of the other tubes. Examine each tube to make certain that no air bubbles cling to the inner surface.

4. Wipe each tube clean with a soft cloth or Kimwipe just before inserting tube into the instrument sample holder.

5. With cover of holder closed, set instrument to a wavelength of 510 nm; use left front knob to set needle to the left zero transmittance line (dark-current adjust-

ment). Make certain that right front knob is turned counterclockwise as far as it will go; then wipe tube containing acid blank and insert tube in the holder, being careful to align mark on the tube exactly with marker on the tube holder. Make this alignment while inserting tube; do not turn tube once it is completely down in the holder. Now use right front knob to adjust the gain until the meter reads zero absorbance (100% transmittance) at the right side of the meter scale.

6. Remove the blank and insert the first tube of cobalt chloride solution, observing all the precautions as for the blank. Record the absorbance for the sample.

7. Remove sample and reinsert blank to check for meter drift, and rezero if necessary.

8. Repeat steps 6 and 7 for the remaining tubes, being sure to keep them in order so that they can be identified later.

9. Select the 10 tubes having the narrowest range of absorbancies. Clean these and wrap in cloth towel. These will be your matched set of tubes. You must take all precautions to avoid marring the surface of them in any way, as this may destroy the matching qualities. In other words, do not clean them with scouring powder, do not bang together in dishpan or sink, do not lay them on a bare counter surface, do not place in wire or metal rack.

10. Repeat this procedure, using the 1-inch tubes, to produce a matched set. If the absorbancies are excessively high with this longer light path, dilute the sample solution with 1% HCl to bring the absorbance back to midscale.

Practice procedures

The following two procedures are included to permit the novice to practice using a simple colored solution ($KMnO_4$) in place of the reagents for specific procedures.

Apparatus

As above with matched tubes

Reagents

Potassium permanganate solution. Dissolve 0.04 g $KMnO_4$ in 1 liter distilled water.

DETERMINATION OF SPECTRAL TRANSMITTANCE CURVE

Procedure

1. Take two clean, dry, 1-cm matched tubes. To the first, add about 5 ml distilled water. To the second, add about 5 ml $KMnO_4$ solution.

2. Set the wavelength scale at 400 nm, and with cover closed, zero left side of scale.

3. Insert blank into holder and adjust with right knob to zero absorbance.

4. Remove blank, insert sample, and record absorbance.

5. Turn down right gain knob, and set wavelength scale to 425 nm. Remove sample, and check left zero with cover closed.

6. Repeat steps 3, 4, and 5 at increments of 25 nm up to 750 nm (change photocell and filter at 625 nm). Plot absorbance against wavelength on linear graph paper. Plot absorbance on the ordinate and wavelength on the abscissa.

7. On the basis of this plot, determine region of rapid change in absorbance with wavelength. Go back and re-measure absorbancies in this region, using 10-nm instead of 25-nm increments. Smooth out your curve on the basis of this better information. Select the best wavelength to determine concentrations of $KMnO_4$ solutions.

DETERMINATION OF A STANDARD CONCENTRATION CURVE

Procedure

1. Use the $KMnO_4$ solution as a 40 mg/liter stock solution, and carefully prepare the following dilute concentrations, using best quantitative technique: 30 mg/liter, 22.5 mg/liter, 15.0 mg/liter, 7.5 mg/liter.

2. Set the spectrophotometer at the wavelength determined to be best in the above procedure; and using

a distilled water blank, determine absorbance for each concentration, including the stock solution.

3. Plot milligrams of $KMnO_4$ per liter against absorbance on linear graph paper.

pH

The pH of a solution is a measure of its hydrogen ion activity and is the logarithm of the reciprocal of the hydrogen ion concentration. Thus it is important to remember that a change of one pH unit represents a tenfold change in hydrogen ion concentration; for example a pH of 6 has 10 times the hydrogen ions of pH 7, and pH 5 has 100 times the hydrogen ions of pH 7.

The pH of most natural waters falls in the range of 4 to 9, and much more often in the range of 6 to 8. In water, deviation from the neutral pH 7 is primarily the result of hydrolysis of salts of strong bases and weak acids or of weak bases and strong acids. However, dissolved gases such as CO_2, H_2S, and NH_3 also have a significant effect. The majority of natural waters have a somewhat alkaline pH due to the presence of carbonate and bicarbonate. The limnologist is interested in pH and its changes, since they may reflect biological activity and changes in natural chemistry of waters, as well as pollution. It is also an important tool for numerous laboratory procedures.

Apparatus

pH meter
Centigrade thermometer
Kimwipes
Beakers

Reagents

pH buffer solution in the range of water sample being analyzed
Distilled water

Procedure for sample collection

The pH of the water may be subject to change during the interval between sampling and determination by reactions such as oxidation or hydrolysis that take place in the bottle. A loss of gases, biological activity, and the absorption of laboratory fumes are all potential sources of change in pH. The sample should be held on ice and analyzed as soon as possible. The analysis is often done in conjunction with the alkalinity determination.

Use of the pH meter

Although meters are simple to use and instructions are available for each make, a few precautions will improve the reliability of the determinations.

1. Because water samples are usually relatively unbuffered solutions as compared with many solutions of the chemical laboratory, the meter is often subject to drift and is slow in coming to a steady reading. Readings should not be taken before equilibrium has been established between the electrode and the water system, as evidenced by lack of drift.

2. Temperature exerts two significant effects on pH measurements. Ionization varies with temperature and is an inherent problem in all pH measurements; thus the temperature is usually reported along with the pH. This ionization effect diminishes with an increase in alkalinity. The second effect lies in the pH electrodes themselves, since the actual potential varies with temperature. An adjustment on the instrument may compensate for this variation; however, it is always best to keep the temperature of the sample as near to the temperature of the buffer as possible, and the difference must not exceed $10° C$.

3. Since the electrodes themselves are the heart of a pH analytical system, care must be taken to see that they are working properly. Care must be taken to avoid scratching the glass electrode against the beaker or wiping with a dirty, gritty cloth or paper. The fiber junc-

tion reference electrode must have a free flow of electrolyte through the fiber. If this becomes plugged through dirt, grease, drying out, or other causes, the flow must be returned to a free state or the electrode discarded. An electrode that does not require regular refilling with KCl electrolyte is suspect.

4. Problems also arise from polarization of the electrode system. To avoid polarization, the electrodes *must never be removed* from any solution when the function switch is in any position other than standby or off.

Procedure

1. Prepare a buffer solution with a pH near the expected pH of the water sample. Buffers may be prepared in the laboratory following the instructions in *Standard Methods* or may be purchased in many forms. Since dilute solutions are less stable, it is best to prepare buffers by dilutions of concentrates following the instructions on the bottle. Place 25 ml buffer solution in a small beaker.

2. Open the vent of the glass reservoir of the fiber junction electrode (usually a plug or sliding sleeve).

3. Rinse the electrodes with distilled water, and blot with soft paper or Kimwipe.

4. Immerse the electrodes in the buffer solution, measure temperature of buffer solution, and set compensator to that temperature.

5. Set function switch to pH, and adjust meter needle to the pH of the buffer. The instrument is now standardized. *Return switch to standby.*

6. Rinse electrodes in distilled water and blot dry.

7. Immerse electrodes in sample, measure temperature of sample, and set compensator to this temperature.

8. Set function switch to pH, and very gently swirl (or use magnetic stirrer at very slow speed) until pH meter needle stops drifting. Read pH from meter scale. *Return switch to standby.* Raise electrodes, rinse with distilled water, and dry. The instrument is now ready for the next sample or for storage.

Storage of the instrument

Close the glass reservoir to minimize electrolyte loss, and leave electrodes immersed in distilled water.

FREE CARBON DIOXIDE

Carbon dioxide is an end product of decomposing bacteria and of the respiratory processes of plants and animals. It is also added to the water by the action of natural or pollution acids on bicarbonates. Accurate analysis for this dissolved gas is difficult. Waters are frequently supersaturated with carbon dioxide. As a result, almost any handling process of the water in collection or analysis will result in a loss of the gas.

Because of this analytical problem, several indirect methods of estimating carbon dioxide concentration have been attempted. These are based on the equilibria of the forms of carbonates in water of known temperature, pH, total mineral content, and total alkalinity. The nomographic method in *Standard Methods* is good but requires quantitative knowledge of total filterable residue. Though not a routine determination of limnological surveys, it is not difficult. A second nomographic method (Moore, 1939) is much simpler, requiring quantitative knowledge of only pH and total alkalinity. Because temperature and total mineral content are not considered, the possibility for error in this method is increased. If, after the Moore nomograph is checked against other methods, it is found to provide adequate results for a given source of water, it provides one of the most rapid methods. A third indirect method is the U. S. Geological Survey (Rainwater and Thatcher, 1960) calculation method. For total solids concentrations of less than 500 mg/liter and in the pH range of 6.0 to 9.0, a simple calculation based on the bicarbonate equilibria and the hydrogen ion concentration is provided. A final method is direct titration. Completion of this titration occurs at pH 8.3 and is determined either with the pH meter or by the development of a pink color, using a phenolphthalein indicator.

Free carbon dioxide by titration

Apparatus

Titration glassware
Daylight or white light source (if using indicators)
pH meter and accessories (if available)

Reagents

Phenolphthalein indicator solution. See following section, Alkalinity.

Standard sodium carbonate titrant, 0.0454 N. Dissolve 0.602 g anhydrous, primary standard grade Na_2CO_3 that has been oven dried at 140° C overnight in freshly boiled and cooled carbon dioxide–free water, and dilute to the mark in a 250-ml volumetric flask. Store in a rubber-stoppered Pyrex bottle. This solution should not be kept for more than 1 to 2 weeks and, for precise work, should be made fresh daily.

Procedure

Most accurate results are obtained by performing the analysis in the field. If returned to the laboratory, the sealed samples must be placed on ice.

Collect the sample by means of water sampler, and discharge the contents at the bottom of a 100-ml Nessler tube, allowing the sample to overflow several volumes; then withdraw the tube while the sample is flowing. Flick the tube to throw off the water down to the 100-ml mark. Add 10 drops of phenolphthalein indicator. If the sample turns pink, free carbon dioxide is absent. If the sample remains colorless, titrate rapidly with standard sodium carbonate titrant, stirring gently with a glass rod until a faint pink color persists for 30 seconds when viewed through the length of the tube over a white surface. It is wise to have a second cylinder of water to which phenol-phthalein has been added, but without titrant, alongside for comparison in determining the faint pink color. For field titration, a 2- to 5-ml serological pipette with small opening may be used as a burette. To check the possi-bility of carbon dioxide loss during the actual process of titrating, a second sample may be obtained; after adding phenolphthalein, rapidly pour in the full amount of titrant used in the first titration. If the sample turns pink, no carbon dioxide was lost; if it remains colorless, titrate with additional titrant until the pink appears. The result of the second titration is to be accepted.

Calculations

$$\text{mg } CO_2 \text{ per liter} = \frac{A \times N \times 22,000}{\text{ml sample}} = \frac{A \times 998.8}{100} = A \times 9.998$$

where

A = ml of titration
N = normality of sodium carbonate

Free carbon dioxide by calculation

The collection and analysis of samples are the same as described in the section on alkalinity. Alkalinity titration and pH determination must be made as soon as and as accurately as possible. Bicarbonate alkalinity is determined as described under alkalinity relationships.

TABLE 2. Value of $1.589 \times 10^6 [H^+]$ for each 0.1 pH unit between pH 6.0 and 9.0

pH	1.589×10^6 $[H^+]$	pH	1.589×10^6 $[H^+]$	pH	1.589×10^6 $[H^+]$
6.0	1.589	7.0	0.159	8.0	.016
6.1	1.262	7.1	.126	8.1	.013
6.2	1.003	7.2	.100	8.2	.010
6.3	.796	7.3	.080	8.3	.008
6.4	.633	7.4	.063	8.4	.006
6.5	.503	7.5	.050	8.5	.005
6.6	.399	7.6	.040	8.6	.004
6.7	.317	7.7	.032	8.7	.003
6.8	.252	7.8	.025	8.8	.003
6.9	.200	7.9	.020	8.9	.002
				9.0	.002

From Rainwater and Thatcher, 1960.

Calculations

Milligrams of carbon dioxide per liter can be calculated in the pH range of 6.0 to 9.0 using the following:

mg CO_2 per liter = $1.589 \times 10^6 \left[H^+ \right] \times$ mg/liter alkalinity as HCO_3

The value of the term $1.589 \times 10^6 \left[H^+ \right]$ is given in Table 2 for each 0.1 pH unit between 6.0 and 9.0.

Free carbon dioxide by Moore nomograph

Using accurate pH and total alkalinity, estimate milligrams of free carbon dioxide per liter from the nomograph (Fig. 11).

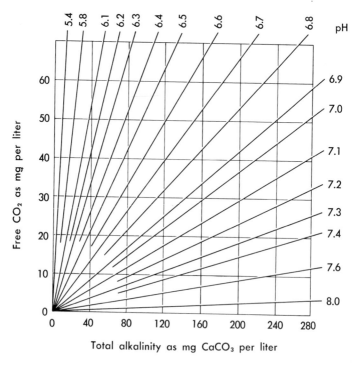

FIG. 11. Moore's nomograph for determination of free carbon dioxide concentration from pH and total alkalinity data.

ALKALINITY

The alkalinity of water is its capacity to accept protons; stated another way, it is the quantity and kinds of compounds present that collectively shift the pH to the alkaline side of neutrality. Although the alkalinity of natural waters is generally the result of bicarbonates, it is usually expressed in terms of calcium carbonate. Three kinds of alkalinity are indicated: hydroxide (OH^-), normal carbonate ($CO_3^=$), and bicarbonate (HCO_3^-); the three are summed as total alkalinity. Carbonates and bicarbonates are common to most waters because carbonate minerals are abundant in nature and because contribution to alkalinity by hydroxides is rare in nature. The presence of hydroxides can usually be attributed to water treatment or to contamination. Expected total alkalinities in nature usually range from 45 to 200 mg/liter.

Principle of determination

There are five conditions of alkalinity possible in the sample: carbonate, bicarbonate, or hydroxide alone, or a combination of carbonate and hydroxide or of carbonate and bicarbonate. Hydroxide and bicarbonate are not found together.

Alkalinity is determined by titrating the water sample with a standard solution of strong acid. The equivalency or end points of the titration are selected as the inflection points in the titration of sodium carbonate with H_2SO_4. These points will vary slightly with temperature, ionic concentration, and free carbon dioxide concentration. This end point may be determined empirically by titration and is that pH where the derivative of $\Delta pH/\Delta ml$ titrant is greatest. However, they are usually taken as pH 8.3 for the carbonate end point and pH 4.5 for the bicarbonate end point. The following reactions occur in titrations (Rainwater and Thatcher, 1960):

$CO_3^= + H^+ \rightarrow HCO_3^-$ (titration to pH 8.3)
HCO_3^- (from $CO_3^=$) $+ H^+ \rightarrow H_2O + CO_2$
HCO_3^- (natural) $+ H^+ \rightarrow H_2O + CO_2$ (titration to pH 4.5)

Apparatus

Titration glassware
Magnetic stirrer if available
Daylight light source

Reagents

Carbon dioxide–free water. Boil for 15 minutes and cool to room temperature; pH must be greater than 6.0.

Phenolphthalein indicator solution. Dissolve 0.5 g phenolphthalein in 50 ml 95% ethyl or isopropyl alcohol. Add 50 ml distilled water, mix, and add weak sodium hydroxide solution drop by drop until a very faint pink color appears.

Stock sulfuric acid solution. Dilute 3 ml concentrated H_2SO_4 to 1 liter; this gives approximately 0.1 N solution.

Standard sulfuric acid solution. Dilute 200 ml 0.1 N stock solution to 1,000 ml with carbon dioxide–free water to give 0.02 N solution. For accurate work this must be standardized against Na_2CO_3 (American Public Health Association, 1971).

Methyl orange indicator solution. Dissolve 0.5 g methyl orange in 1 liter of distilled water.

Bromcresol green–methyl red indicator solution. Dissolve 0.02 g methyl red and 0.10 bromcresol green in 100 ml 95% alcohol.

Procedure

Collect 250-ml or greater samples without aeration in glass or polyethylene bottles. Alkalinity is susceptible to change between time of collection and analysis. The main problem is loss of carbon dioxide from solution, which results in the conversion of bicarbonate to carbonate. Total alkalinity values are probably somewhat more stable than the relative values of the common alkalinity components. Analysis must be done within a few hours of collection, and the sample must be kept sealed until analysis is ready to begin; it must not be filtered, diluted, or altered in any way. Gently pour out 100 ml of sample and place in a 250-ml Erlenmeyer flask (if indicators are to be used) or a 200- to 250-ml beaker (if pH meter is to be used).

Titration with pH meter. Place the prepared sample on a magnetic stirrer and add spin bar. Place prepared pH electrodes in sample. Turn on stirrer at slow speed (do not agitate surface of liquid), and record sample pH when meter drift stops.

Titrate with standard acid solution, recording titrant volume at pH 8.3 and pH 4.5. If a magnetic stirrer is not available, swirl the sample gently by hand.

Titration with indicators. Although a meter is preferable, indicators for determination of pH end points may be used. Add 4 drops of phenolphthalein indicator to the sample. If a pink color appears, hydroxide or normal carbonate is present. Titrate with standard sulfuric acid until color disappears, and record milliliters of acid used. If no pink color appears, or after phenolphthalein titration, add 2 drops of methyl orange indicator, and titrate to a faint orange. This is a difficult end point and usually requires some experience to recognize. Overtitration (to pH 4 or below) produces a pink color. As an alternate method, one can substitute bromcresol green–methyl red indicator for the methyl orange and titrate until the blue changes to a *clear* pink (a gray-pink appears before the final change). It is wise to practice recognizing this end point by using meter and indicator together.

Calculations

Phenolphthalein alkalinity

$$\text{as mg CaCO}_3 \text{ per liter} = \frac{A \times N \times 50{,}000}{\text{ml sample}} = \frac{A \times 1{,}000}{100} = A \times 10$$

Total alkalinity

$$\text{as mg CaCO}_3 \text{ per liter} = \frac{B \times N \times 50{,}000}{\text{ml sample}} = \frac{B \times 1{,}000}{100} = B \times 10$$

where

A = ml titration to pH 8.3 (phenolphthalein end point)
B = ml total titration from start to pH 4.5 (methyl orange end point)
N = normality of acid

Calculation of alkalinity relationships

Results obtained from the titrations offer a means of classifying the three principal forms of alkalinity present. Mathematical conversion of titration results may be made using Table 3.

TABLE 3. Alkalinity relationships

Result of titration	OH alkalinity as $CaCO_3$	CO_3 alkalinity as $CaCO_3$	HCO_3 alkalinity as $CaCO_3$
P = 0	0	0	T
P < 1/2 T	0	2P	T − 2P
P = 1/2 T	0	2P	0
P > 1/2 T	2P − T	2(T − P)	0
P = T	T	0	0

P, phenolphthalein titration; T, total titration.

HARDNESS

Water that contains salts of calcium and magnesium and, to a lesser extent, other polyvalent metals, such as iron, aluminum, and manganese, and that requires large amounts of soap to lather or that, upon evaporation, forms a deposit on the container is referred to as hard water. In early methodology, hardness was measured as the capacity of water to precipitate soap when a liquid soap solution was shaken with the water sample to form a lather persisting for 5 minutes. Hardness is now best defined as a characteristic of water representing the total concentration of calcium and magnesium ions expressed as milligrams of $CaCO_3$ per liter. When other ions are present in insignificant amounts, the hardness will be equal to or less than the sum of the carbonate and bicarbonate alkalinities and is termed carbonate hardness. If the hardness exceeds the sum of these alkalinities, the presence of other ions is indicated, and the excess is expressed as noncarbonate hardness. Hardness determination is thus corroboratory to the findings of alkalinity; but when used in conjunction with calcium determinations, it allows for the calculation of magnesium concentration, circumventing the laboratory analysis for this ion. Hard water is not generally considered harmful to man (except as a possible cause of kidney stones); and within limitations, the ions present are necessary for normal plant and animal survival and growth. Evidence suggests that hardness may affect tolerance of fishes to toxic metals.

The Eriochrome Black T indicator used in this titration produces a red color in the presence of calcium and magnesium ions. The disodium salt of EDTA forms a stable, colorless complex with these ions, effectively removing them from solution. Thus by titration of the sample containing the indicator with EDTA, the calcium and magnesium ions are quantitatively removed. When all ions are removed, the indicator changes to a bright blue color.

Apparatus

Titration glassware
Magnetic stirrer if available
Daylight light source

Reagents

Buffer solution. Use commercially prepared buffer as directed, or prepare per instructions in *Standard Methods.*

Indicator. Mix 0.8 g Eriochrome Black T dye and 100 g NaCl to prepare a dry powder mixture.

Standard EDTA titrant, 0.01 M. Dissolve 0.3723 g Na_2EDTA-dihydrate in distilled water and dilute to 100 ml.

Check by titrating against a standard calcium solution: 1.00 ml = 1.00 mg $CaCO_3$ = 0.4008 mg Ca.

Standard calcium solution. Weigh 1.000 g anhydrous calcium carbonate powder, primary standard grade, into a 500-ml Erlenmeyer flask. Add slowly one volume HCl diluted with an equal volume of distilled water (1+1 HCl) until all the $CaCO_3$ has dissolved. Add 200 ml distilled water and boil for a few minutes to expel CO_2. Cool, and adjust to pH 5.0 with either NH_4OH (about 3 N) or 1+1 HCl. Transfer to a 1-liter volumetric flask, washing out the Erlenmeyer flask several times with distilled water and adding to volumetric flask. Then dilute to mark with distilled water. (1.00 ml = 1.00 mg $CaCO_3$ = 0.4008 mg Ca.)

Procedure

1. Dilute 25 ml of sample to about 50 ml with distilled water in titration flask.

2. Add 1 to 2 ml of buffer solution to bring pH to 10.0 or 10.1.

3. Add approximately 0.1 g indicator powder (just a bit on the tip of a spoon or spatula).

4. Titrate with EDTA over a white surface with daylight or white light. Stir continuously until the last red tinge disappears. Add the last drops slowly, allowing about 5 seconds between drops. The entire duration of titration should not exceed 5 minutes and should not require more than 15 ml of titrant. If more titrant than this is used, take a smaller aliquot and repeat titration. An indistinct end point suggests interference and calls for an inhibitor after step 2. Faulty (old) indicator powder also produces an indistinct end point.

Calculations

$$\text{EDTA hardness as mg } CaCO_3 \text{ per liter} = \frac{A \times B \times 1{,}000}{\text{ml of sample}}$$

where

A = ml titration
B = mg $CaCO_3$ equivalent to 1.00 ml EDTA titrant

Express results as either total hardness, carbonate hardness, or noncarbonate hardness.

CALCIUM

Calcium can be leached from practically all rocks but is much more prevalent in waters from regions with deposits of limestone, dolomite, and gypsum. Regions where granite or silicious sand predominate have very low calcium levels in the waters. Concentrations in waters from limestone areas range from about 30 to 100 mg/liter. Calcium is important to the biological productivity of waters. Waters with a concentration of 10 mg or less per liter are usually oligotrophic, while waters with 25 mg or more per liter are usually distinctly eutrophic.

The EDTA titrant used in this calcium determination is the same as for water hardness, and reacts with both calcium and magnesium. To determine calcium only, the pH is made sufficiently high so that the magnesium precipitates as magnesium hydroxide, and an indicator specific for calcium is used.

Apparatus

Titration glassware
Magnetic stirrer if available
Daylight or white light source

Reagents

Sodium hydroxide, 1 N. Dissolve 4 g NaOH in distilled water, and when cool, dilute to 100 ml.

Murexide indicator. Grind together in a mortar 0.2 g powdered dye and 100 g NaCl. Store in tightly stoppered bottle.

Standard EDTA titrant, 0.01 M. Same as in hardness determination. (1 ml = 0.4008 mg Ca.)

Procedure

1. Take a sample that contains less than 10 mg Ca. Usually a 50-ml water sample is correct, but if total alka-

linity is greater than 250 mg/liter it probably will be better to take a smaller aliquot and dilute to 50 ml with distilled water.

2. Add 1 to 2 ml NaOH solution to produce a pH of 13 to 14. Stir.

3. Add about 0.2 g indicator powder. The color change is from pink to purple upon titration.

4. With continuous stirring, titrate slowly over a white surface with the EDTA titrant. This end point recognition is facilitated by preparing a reference end point by adding NaOH, indicator, and 1 or 2 ml EDTA to 50 ml distilled water.

Calculations

$$\text{mg Ca per liter} = \frac{A \times B \times 400.8}{\text{ml of sample}}$$

where

A = ml titration for sample
B = mg $CaCO_3$ equivalent to 1.00 ml EDTA titrant

With the correctly standardized titrant and a 50-ml sample, the calculation is:

$$\text{mg Ca per liter} = A \times 8.016$$

MAGNESIUM

Magnesium in natural waters comes mainly from the leaching of igneous and carbonate rocks. In areas where these sources are common, magnesium concentrations in water often range from 5 to 50 mg/liter. Magnesium is related to water hardness in the same manner as calcium. This element is of concern to the limnologist in its role as an essential nutrient in plant growth and development, especially as related to its function in the chlorophyll molecule.

Procedure

There are several analytical methods for magnesium, but if both calcium concentration and hardness are known, magnesium concentration may be calculated from them (Rainwater and Thatcher, 1960).

Milliequivalents of hardness per liter are calculated from milligrams of hardness per liter. The milliequivalents of calcium per liter are subtracted from this, and the difference is multiplied by the equivalent weight of magnesium. This gives magnesium in milligrams per liter.

Calculations

mEq hardness per liter = mg hardness per liter × 0.01998
mEq Ca^{+2} per liter = mg Ca^{+2} per liter × 0.0499
mg Mg^{+2} per liter = 12.16 × (mEq hardness per liter − mEq Ca^{+2} per liter)

DISSOLVED OXYGEN

The dissolved oxygen (DO) content of waters results from (1) the photosynthetic and respiratory activities of the biota in the open water, the benthos, and the aufwuchs; and (2) the diffusion gradient at the air-water interface and distribution by wind-driven mixing. Generally 3 mg DO or less per liter is considered stressful to most aquatic vertebrates.

The azide modification of the Winkler method is applicable to most natural waters; however, nitrites, ferrous and ferric iron, sulfites, sulfides, polythionates, and other sulfur compounds interfere with the correct determination of DO when the Winkler reagents are used. Both nitrites and ferric iron liberate additional iodine from the potassium iodide during the course of the procedure so that the final titration is too high. Ferrous iron, sulfites, and organic substances remove some of the iodine liberated from the potassium iodide by the manganic compounds so that the final titration is too low. If these interferences are suspected, orientation tests to demonstrate their presence or absence may be conducted (Ellis and others, 1948). Any readily oxidizable or reducible substances may cause interference. Oxidizable substances cause low DO results, while reducible substances cause high results. In this method, the sodium azide eliminates the interference of nitrite, and the addition of a simple initial potassium fluoride step overcomes the effect of ferric iron up to 200 mg iron per liter. In the absence of interference, this method is reproducible to $\pm$ 0.01 except at low concentrations.

Principle of determination

This method depends on the formation of a loose precipitate (floc) of manganous hydroxide. The oxygen dissolved in the water is rapidly absorbed by the manganous hydroxide, forming a higher oxide, which may be in the following form:

$$MnSO_4 + 2KOH \rightarrow Mn(OH)_2 + K_2SO_4$$
$$2Mn(OH)_2 + O_2 \rightarrow 2MnO(OH)_2$$

As it settles, the $Mn(OH)_2$ floc acts as a "gathering" agent for the DO. Upon acidification in the presence of iodide, iodine is released in a quantity equivalent to the DO present:

$$MnO(OH)_2 + 2KI + H_2O \rightarrow Mn(OH)_2 + I_2 + 2KOH$$

The liberated iodine is then titrated with a standard sodium thiosulfate solution, with starch used as the indicator (Rainwater and Thatcher, 1960):

$$I_2 + 2S_2O_3^{-2} \rightarrow S_4O_6^{-2} + 2I^{-1}$$

Reagents

Manganese sulfate solution. Dissolve 120 g $MnSO_4 \cdot 4H_2O$, or 100 g $MnSO_4 \cdot 2H_2O$, or 91 g $MnSO_4 \cdot H_2O$ in distilled water; filter, and dilute to 250 ml (do not store in glass-stoppered bottle).

Alkali-iodine-azide reagent. Dissolve 125 g sodium hydroxide and 33.75 g sodium iodide in distilled water, and dilute to 250 ml. To this solution add 2.5 g sodium azide *(poison)* that has been dissolved in 10 ml distilled water.

Starch solution. Prepare an emulsion of 6 g starch in a

beaker with a *small* amount of distilled water. Pour this emulsion in 1 liter of distilled water, boil for a few minutes, and let settle overnight. Save the clear supernate, and preserve by adding 1.25 g salicylic acid (it is best to store the stock bottle under refrigeration).

Sulfuric acid. Concentrated.

Sodium thiosulfate stock solution, 0.10 N. Dissolve 12.41 g $Na_2S_2O_3 \cdot 5H_2O$ in boiled and cooled distilled water; dilute to 500 ml. Preserve by adding 2 g borax ($Na_2B_4O_7 \cdot 10H_2O$), or 0.5 g NaOH, or 2.5 ml chloroform.

Standard sodium thiosulfate titrant, 0.0125 N. Dilute 125.0 ml sodium thiosulfate stock solution, 0.10 N, to 1,000 ml with distilled water. Preserve by adding either 3 g borax, or 0.4 g NaOH, or 5 ml chloroform. This solution must be standardized by comparison with standard potassium iodate (KIO_3). Place 25.00 ml of the standard KIO_3 in a 250-ml Erlenmeyer flask, and add 75 ml demineralized water. Dissolve 2 g potassium iodide (KI) in this solution. After all of the potassium iodide is dissolved, add 10 ml of 3.6 M H_2SO_4 (carefully dilute 200 ml concentrated H_2SO_4 to 1 liter). Allow this to stand at room temperature in the dark for 2 to 5 minutes. Titrate with the sodium thiosulfate, adding 2 ml starch indicator when the titration has reached a pale straw color. Continue titration until blue color disappears. The actual normality of the standard sodium thiosulfate may be calculated as follows:

$$N \text{ of } Na_2S_2O_3 = \frac{25}{ml \ Na_2S_2O_3 \ used \ in \ titration} \times 0.0125$$

The standard sodium thiosulfate may then be adjusted to exactly a 1:1 equivalence by the addition of either stock thiosulfate solution or distilled water. An alternative is to consider the result of the standardization as a correction factor (K) and correct the empirically determined DO values by multiplying by K.

Standard potassium iodate solution, 0.0125 N. Although it is probably best for most inexperienced workers to purchase stock potassium iodate solutions of 0.10 N and make the appropriate dilutions, it is possible to prepare a standard iodate solution as follows. Oven dry approximately 1 g KIO_3 for 1 hour at 105° C. Dissolve 0.4459 g in demineralized water, and dilute to 1,000 ml. This standard potassium iodate solution must be made up fresh daily.

Procedure

1. Insert the delivery tube of the water sampler to the bottom of a glass-stoppered bottle. (The 300-ml BOD bottles with tapered glass stoppers are preferred, but any 125-ml or 250- to 300-ml bottle may be used.) Allow the water to flow into the bottom of the bottle until it has overflowed 2 or 3 times its volume, making sure that no air bubbles are retained. With flow continuing, remove the tube from the bottle, and immediately stopper without trapping any air bubbles. (This is easy with BOD bottles, but more difficult with flat stoppers.) Using flat stoppers, tip the bottle slightly, partially insert the stopper, and then with a quick twisting motion, seat the stopper.

2. If percent saturation is to be computed, water temperature must be taken.

3. Add 2 ml $MnSO_4$ solution to the sample in a 300-ml bottle by inserting tip of pipette just below water surface.

4. Add 2 ml alkaline-iodide-azide reagent in same manner as for $MnSO_4$.

5. Stopper as described above, and mix by inverting 10 to 20 times. Allow the floc to settle until one third of the bottle is clear, then mix again. Allow the floc to settle a second time.

6. When at least one third of the bottle is clear, carefully remove the stopper and add 2 ml concentrated H_2SO_4 by allowing the acid to flow down the neck of the bottle. Restopper and invert gently until solution is complete. In the absence of interferences, the treated sample may be held for up to 3 days under refrigeration before titration.

7. Titration: mix by inversion and pour out 100 ml of sample for titration. For highest precision, this volume taken for titration should correspond to 100 ml of sample water before the addition of reagents. Thus the volume taken for titration using the quantities of reagents described above is 101.5 ml. (For method of calculation see American Public Health Association, 1971, or Ellis and others, 1948.) For routine work this correction is not necessary.

8. Titrate with 0.0125 N thiosulfate solution to a very pale yellow color, add 1 ml starch solution (if the sample turns black instead of blue, you added the starch too soon; this may result in slight error), and continue titration until the first disappearance of the blue color. Record milliliters of thiosulfate. Disregard any reappearance of

TABLE 4. Solubility of oxygen in pure water exposed to water-saturated air at mean sea level pressure of 760 mm Hg

Temperature (° C)	DO (mg/liter)	Temperature (° C)	DO (mg/liter)
0	14.16	18	9.18
1	13.77	19	9.01
2	13.40	20	8.84
3	13.05	21	8.68
4	12.70	22	8.53
5	12.37	23	8.38
6	12.06	24	8.25
7	11.76	25	8.11
8	11.47	26	7.99
9	11.19	27	7.86
10	10.92	28	7.75
11	10.67	29	7.64
12	10.43	30	7.53
13	10.20	31	7.42
14	9.98	32	7.32
15	9.76	33	7.22
16	9.56	34	7.13
17	9.37	35	7.04

Modified from Hutchinson, 1957.

blue color. If the color does reappear within 1 to 2 minutes, it is an indication of a very accurate titration, for if the end point is much overrun, the recoloration does not occur. If the end point is overrun, take an additional 10 ml of sample and add to the first 100 ml; this should cause a blue color to reappear. Then titrate to end point and mathematically correct for the larger sample.

Calculations

Since 1 ml of 0.0125 N thiosulfate is equivalent to 0.1 mg DO, each milliliter of sodium thiosulfate titrant used is equivalent to 1 mg DO per liter when a volume of 100 ml is titrated.

To express the results as percent saturation, the solubility data from Table 4 may be used.

Lower solubilities are found in natural waters because of dissolved salts, but these differences are not significant in nonpolluted fresh waters. To calculate the solubility of DO at any other atmospheric pressure, the following formula may be used, provided the water temperature is below 25° C.

$$S' = S \frac{P}{760}$$

where

　　S' = solubility in mg/liter
　　S　= solubility at 760 mm Hg (760 mm = 29.92 inches)
　　P　= atmospheric pressure at time of sample collection

ORTHOPHOSPHATE

Phosphorus in waters is present in several soluble and particulate forms, including organically bound phosphorus, inorganic polyphosphates, and inorganic orthophosphates. These orthophosphates are usually ions of phosphoric acid. At pH concentrations of most natural waters (less than pH 9.0), the dihydrogen and monohydrogen phosphate ions are prevalent, although analytical techniques usually employed do not distinguish between ionic states and all inorganic phosphate is usually considered as PO_4.

Because phosphorus is a biologically active element, it cycles through many states in the aquatic ecosystem, and its concentration in any one state depends on the degree of metabolic synthesis or decomposition occurring in that system; for example, lower concentrations are expected at times of high synthetic activity. Although the natural source of phosphorus to waters is from leaching of phosphate-bearing rocks and from organic matter decomposition, additional sources are found in manmade fertilizers, domestic sewage, and detergents. Phosphorus is lost from the water by chemical precipitation to the sediments and by adsorption on clays.

The following method of Murphy and Riley (1962) specifically measures the orthophosphate concentration. If either polyphosphates or organic phosphorus determinations are desired, these may be made by differential acid hydrolysis to the orthophosphate state and then this method employed. Details concerning hydrolysis are available in the handbook by Strickland and Parsons (1968).

The Murphy and Riley method is usually preferred because of its adequate sensitivity and lack of interferences. The stannous chloride procedure (American Public Health Association, 1971) is very sensitive, but suffers from severe arsenic interference. Minimum detectable concentration by the Murphy and Riley method is approximately 5 μg/liter (parts per billion) when a 1-inch sample tube is employed with a Spectronic 20. For relatively unpolluted waters in many portions of the United States, this is inadequate sensitivity. In this case, an additional alcoholic extraction step (Stephens, 1963) may be employed to bring the sensitivity to the 1-ppb range.

It must be stressed that when one is analyzing for phosphorus in the ppb range, laboratory contamination from dust and detergents will often produce higher concentrations than the lake or river water itself. Phosphorus-containing detergents should be banned from the laboratory in which phosphorus analyses are made. All glassware to be used for phosphorus determination must be acid-washed and preferably stored submerged in a dilute acid (1% to 5%) solution. This glassware should never be used for any other types of laboratory analysis. Reagent blanks on glass-redistilled water should be carried through with each analysis to confirm the lack of contamination.

Because of the reported possibility of phosphorus adsorption onto polyethylene, samples for phosphorus analysis should be collected in acid-washed Pyrex bottles. Samples should be refrigerated immediately, and analysis should be completed within a few hours. If analysis must be delayed, any filtration must be done immediately, and the samples must be stored by freezing.

Apparatus

Spectronic 20 with red phototube and filter set at 880 nm (690 nm may also be used but tends to be less sensitive)

1-inch matched test tubes

Acid-washed glassware (Erlenmeyer flasks, graduates, pipettes)

Reagents

Glass-distilled water. Use only glass-redistilled water stored in polyethylene bottles for all reagents and blanks.

Phenolphthalein indicator. See section on alkalinity.

Sulfuric acid solution. Slowly add 7 ml concentrated H_2SO_4 to distilled water, and dilute to 50 ml.

Ammonium molybdate reagent. Dissolve 4.0 g $(NH_4)_6Mo_7O_{24} \cdot 4H_2O$ in 100 ml distilled water. Refrigerate in polyethylene bottle for storage.

Potassium antimonyl tartrate. Dissolve 1.3715 g potassium antimonyl tartrate ($K[SbO]C_4H_4O_6 \cdot 0.5 \ H_2O$) in 400 ml distilled water, and make to volume in a 500-ml volumetric flask. Refrigerate in a polyethylene bottle for storage.

Ascorbic acid solution. Dissolve 1.76 g ascorbic acid in 100 ml distilled water. Freeze in a polyethylene bottle for storage. This may be thawed and refrozen repeatedly.

Strong acid solution. Slowly add 30 ml concentrated H_2SO_4 to 60 ml distilled water.

Mixed reagent. Add 5 ml potassium antimonyl tartrate solution to 50 ml sulfuric acid solution and mix. Add 15 ml ammonium molybdate solution and mix. Add 30 ml ascorbic acid solution and mix. This reagent should be prepared fresh for each set of analyses.

Stock phosphorus solution. Although it is preferable for the occasional worker to purchase commercially prepared stock solutions, stock and standard solutions may be prepared as follows. A stock solution where 1.00 ml equals 0.05 mg phosphorus is prepared by dissolving 0.2197 g potassium dihydrogen phosphate (KH_2PO_4) (oven dried at 105° C for 1 hour) in distilled water. Dilute this to 1.0 liter. Add 1 ml chloroform and store in the dark under refrigeration. This stock solution will remain stable for months.

Standard solution (1.00 ml $= 0.5 \mu g$ P). Dilute 1.0 ml of stock phosphorus solution to 1.0 liter with glass-distilled water. This standard solution should not be stored for more than a few days.

Procedure

Water should be settled free of suspended matter, or it may be filtered, if at all turbid, using either membrane, glass fiber, or paper filters that have been shown to be free of phosphorus. Rinse filter with acid rinse, and follow with glass-distilled water rinse before filtering sample. A blank of glass-distilled water should be run through the same procedure so that contamination at any step of the procedure will show up as a blue color in the blank.

Carry 2 different standards of approximately the expected phosphate concentration through with the samples.

1. Add 1 drop of phenolphthalein indicator to a 50-ml

sample. If a pink color appears, add strong acid solution drop by drop until the color disappears; do not add more than 5 drops of acid. If more than 5 drops are required, take a 40-ml sample, discharge pink color with acid; then make up to 100 ml with distilled water, and make the necessary correction in your calculations for the smaller volume of sample.

2. Add 8.0 ml of the mixed reagent to the sample, and mix thoroughly. After 10 minutes measure the absorbance of each sample at 880 or 690 nm in the spectrophotometer, using the reagent blank as the reference solution. Determine concentration of phosphorus in this sample from a standard phosphorus calibration curve. If the two standards run with the samples do not agree with this standard curve, the reagents have changed and a new curve must be made.

Calculations

$$\text{mg P per liter} = \frac{\text{mg P from curve} \times 1{,}000}{\text{ml sample}}$$

$$\text{mg PO}_4 \text{ per liter} = \text{mg P per liter} \times 3.067$$

Standard curve for phosphorus. Standard phosphorus solution: 1.00 ml = 0.5 μg P.

For many natural waters, the dilutions shown in Table 5 have been found suitable.

TABLE 5. Volume of standard solution, to be made up to 50 ml with distilled water, and the resultant concentrations

Milliliters of standard solution	Phosphorus concentration (μg per liter)
0.5	5
1.0	10
3.0	30
5.0	50
10.0	100

NITRATE AND NITRITE NITROGEN

Nitrate nitrogen is the most highly oxidized state of the element found in water. In most natural waters it is also the commonest state. It is brought into the aquatic system by the bacterial oxidation of atmospheric nitrogen and by decomposition of organic matter in the watershed. Once in the water, it represents both the beginning and the end of the nitrogen cycle because it is used directly in the assimilatory activities of the plants and it represents the end product of aerobic decomposition of organic nitrogen–containing molecules.

A simple yet sensitive and widely accepted method for the determination of nitrate nitrogen is not available. The phenoldisulfonic acid method (American Public Health Association, 1971) is the most sensitive method in common use, but suffers from severe interferences that must be removed by time-consuming techniques. The Brucine method (American Public Health Association, 1971) is less sensitive and requires extremely careful temperature control. The following cadmium reduction method, using prepared Hach reagents, is adequate for most survey work. This method measures both nitrate and nitrite nitrogen. Since nitrite nitrogen is rarely measurable in most natural water, a preliminary test to demonstrate its absence allows one to proceed with the cadmium reduction analysis, assuming that all color development is due to nitrate nitrogen.

Apparatus

Spectronic 20 with blue tube and set at 525 nm
Matched 1-inch test tubes
Glassware

Reagents

Hach NitraVer IV powder pillows
Hach NitriVer powder pillows
Stock nitrate solution (100 mg N per liter). Dissolve 0.7218 g anhydrous potassium nitrate (KNO_3) and dilute to 1,000 ml with demineralized water.

Standard nitrate solution (1 mg N per liter). Dilute 10 ml stock solution to 1,000 ml with demineralized water. Prepare fresh weekly.

Procedure

1. Add the contents of one NitriVer powder pillow to 25 ml of sample in an Erlenmyer flask. Wait 15 minutes after mixing. If a color develops, nitrite nitrogen is present. If nitrite is absent, continue with the procedure for the determination for nitrate nitrogen.

2. Place 25 ml of sample in an Erlenmeyer flask, add the contents of one NitraVer IV powder pillow to the flask, and shake vigorously for 1 minute. If nitrate is present, a red color will develop. Wait 3 minutes for full color development.

3. Measure absorbance using the 1-inch test tubes and the Spectronic 20 set at 525 nm. Determine milligrams of nitrogen from a standard curve.

SILICA

Silica is extremely common in nature as a constituent of igneous rocks, quartz, and sand. Natural waters often contain 1 to 10 mg silica per liter and very rarely exceed 60 mg silica per liter. The chemistry of silica in solution is not known with certainty; however, most silica is present in the nonionized form. Silica is primarily of concern to the limnologist as the principal component of the cell wall of diatoms. A silica cycle occurs in bodies of water that contain diatoms. Silica is removed from the water during synthesis and slowly returned to the water by re-solution of the dead organisms.

Apparatus

Spectronic 20 with blue phototube, wavelength at 410 nm
1-cm matched test tubes
Pyrex glassware

Reagents

Distilled water. Use deionized water stored in polyethylene bottles for all reagents.

Hydrochloric acid. 1 + 1 vv.

Ammonium molybdate reagent. Dissolve 5 g $(NH_4)_6Mo_{24} \cdot 4H_2O$ in distilled water with slow warming; then dilute to 50 ml. Filter if any insoluble material persists. Adjust the pH of this solution to 7 to 8 with silica-free ammonium or sodium hydroxide, and store in a polyethylene bottle.

Oxalic acid solution. Dissolve 5 g $H_2C_2O_4 \cdot 2H_2O$ in distilled water and dilute to 50 ml.

Reducing agent. Dissolve 0.5 g of 1-amino-2-naphthol-4-sulfonic acid and 1 g anhydrous sodium sulfite (Na_2SO_3) in 50 ml distilled water (warm if necessary). Add this to a solution of 30 g sodium bisulfite ($NaHSO_3$) dissolved in 150 ml distilled water. Mix and filter into a polyethylene bottle. Store this reducing agent in the refrigerator. When the reagent becomes dark colored, it is no longer usable and must be discarded.

Procedure

Samples are to be stored in polyethylene bottles and may be kept under refrigeration for up to 1 week. Excessive silica concentrations are reported when samples are held for a period of weeks prior to analysis, during which time diatom populations decompose. In addition to possible contamination by glassware, other sources of error are the presence of phosphate or turbidity or both in the water. Both phosphorus and silica react with the molybdate to form acids. However, the addition of oxalic acid destroys the molybdophosphoric acid, leaving only the molybdosilicic acid, which produces a yellow color proportional in intensity to the concentration of the silica. Turbidity, if obvious, must be removed by filtration through any filtration apparatus shown to be free of silica. The reagent blank must also be run through the same procedures, including filtration, so that it will detect and compensate for any contamination.

It is possible to compensate photometrically for turbidity, thus canceling its effect. This technique may be used in the silica procedure. Compensation is accomplished as follows: two identical aliquots of each sample are prepared. All of the reagents are added to one aliquot as described below; all reagents *except* the molybdate are added to the other aliquot in normal sequence. The aliquot without the molybdate is used to set the photometer to zero absorbance; then the corresponding aliquot with the molybdate is placed in the light path, and absorbance is read. This automatically corrects for turbidity present. Even if compensation is used, a distilled water reagent blank must also be run to check for contamination. Filtration or compensation is not necessary unless the sample is obviously turbid.

This procedure measures only the molybdate-reactive silica. Certain forms of silica do not react with molybdate. The biological significance of these nonreactive forms is unknown. An optional pretreatment to convert any nonreactive silica to the reactive form is sometimes used and is described in *Standard Methods*. Since the procedure described below omits this conversion step, results should be reported as molybdate-reactive silica.

1. To a 50-ml sample, add in quick succession 1.0 ml 1+1 HCl and 2.0 ml ammonium molybdate reagent, and mix by swirling.

2. Allow treated sample to stand for 5 to 10 minutes.

3. Add 1.5 ml oxalic acid solution and mix well.

4. After 2 minutes, add 2.0 ml of reducing agent and mix. Read the color developed after 5 minutes, using 1-cm matched tubes, with Spectronic 20 set at 815 nm. A less sensitive wavelength at 650 nm may be preferred for high silica concentration.

5. Determine mg SiO_2 from standard curve prepared

from dilutions of silica standard solution (1 ml = 0.010 mg SiO_2).

Calculations

$$\text{mg } SiO_2 \text{ per liter} = \frac{\text{mg } SiO_2 \times 1,000}{\text{ml sample}}$$

SULFATE

In many natural waters, sulfate is the second most common anion, being derived from most sedimentary rocks. In lakes, sulfur is cyclic and involves organically reduced forms as well as the common free $SO_4^=$ ion. Although few studies have been conducted on sulfur metabolism in algae, sulfate is the common form taken up by higher plants and is probably of equal importance to algae. The turbidometric method described here is one of the least precise; but because of its simplicity, it is often used for routine work. As described here, the method is basically that of *Standard Methods* but makes use of the convenient Hach reagents.

Apparatus

Spectronic 20 with blue phototube set at 420 nm
Matched 1-inch test tubes
Glassware

Reagents

Standard sulfate solution (1.00 ml = 0.01 mg SO_4). Using a microburette, measure 10.41 ml of standard 0.0200 N H_2SO_4 titrant (from alkalinity procedure) into a 100-ml volumetric flask, and dilute to mark with distilled water.

SulfaVer powder. From Hach Chemical Co.

Procedure

Carry a distilled-water blank through all steps.

1. To a 25-ml sample in a 125-ml flask add 1.0 g SulfaVer powder, and swirl evenly for 1 minute. A suspension of barium sulfate forms.

2. Pour entire sample into a 1-inch matched test tube, and let stand for 3 minutes.

3. Read absorbance produced by this suspended turbidity at a wavelength of 420 nm. Estimate milligrams of sulfate by comparing with a standard curve prepared by running known standard concentrations through the same procedure, *being careful that all timing is the same*. The highest standard should not exceed that equivalent to 40 mg/liter (1 mg/25 ml sample), since this method fails above that concentration.

Calculations

$$\text{mg } SO_4 \text{ per liter} = \frac{\text{mg } SO_4 \text{ in sample} \times 1,000}{\text{ml of sample}} = \text{mg } SO_4 \text{ in sample} \times 40$$

SPECIFIC CONDUCTANCE

Specific conductance is a measure of a water's capacity to conduct an electric current. It is the reciprocal of resistance for which the standard unit is an ohm. Since conductance is the inverse of resistance, the unit of conductance is the mho, or in low-conductivity natural waters, the micromho. Because the measurement is made using two electrodes placed 1 cm apart, specific conductance is generally reported as micromhos per centimeter. The relationship of specific conductance to ionized matter concentration varies with both the quality and quantity of the ions present. However, at low concentrations the ions move and behave independently, and the conductance-concentration relationship is almost linear. Temperature of the solution affects the ionic velocity, and thus, the specific conductance. In unpolluted waters, conductance increases from 2% to 3% per degree centigrade, and generally a correction of 2.5% is usable. Since temperature is an integral part of these data, it must always be reported along with the specific conductance. For convenience, most literature values are found corrected to one standard temperature. Sanitary engineers and geochemists have long used 25° C as a standard temperature. Recently limnologists have used 18° C as standard, since this is closer to the average natural water temperature found in the United States.

The greatest value of specific conductance to the limnologist is the estimation, by this very simple procedure, of the total concentration of dissolved ionic matter in the water, which in turn is related to water fertility. Total dissolved matter may be estimated from conductance data by multiplying by some empirical factor, which usually varies from 0.5 to 1.0. (This method is considered under the procedure for dissolved matter.) Specific conductance values also provide a quick check of alteration of the total quality of water due to the addition of many pollutants. With experience, the analyst

FIG. 12. Materials needed to determine specific conductance include an electric bridge (meter) and dip cell, one container for the water sample and a second for distilled water, and a thermometer.

may also use conductance data to estimate the aliquots of samples to be taken for many chemical analyses so that concentrations fall within the range of the method.

Apparatus

Conductivity meter with dip cell
Water bath at 18° C or 25° C, if conductivity meter
    lacks temperature compensation
Thermometer
2 beakers of 200- to 250-ml capacity or similar vessel
    deep enough to cover vertical dip cell

Procedure

1. Connect lugs of dip cell to instrument (the lugs may be connected to either post). A cell constant of 1 is best for most natural waters; other cells or ranges are available for weak or strong conductors. If the cell constant is unknown or if it is suspected that through time the constant has changed, the actual cell constant may be calculated as follows: prepare a 0.00702 N potassium chloride solution. (Dissolve 0.5232 g KCl, oven dried at 180° C for 1 hour, in demineralized water, and dilute to 1,000 ml.) This electrolyte has a specific conductance of 0.001000 mhos at 25° C. The cell constant is calculated by multiplying the measured resistance of this electrolyte solution at 25° C by the known specific conductance.

2. Fill 2 beakers with water sample to a depth sufficient to cover the vent holes in the dip cell when immersed. Place in water bath, and wait for equilibrium if necessary.

3. Measure temperature of the second beaker, and set temperature compensator on instrument to that temperature.

4. Rinse dip cell thoroughly in the first beaker, and then transfer dip cell to second beaker. Turn on instrument, and while gently moving the dip cell, keeping

the holes submerged all the time, take a reading of specific conductance.

5. Turn off instrument, remove and disconnect dip cell, rinse with distilled water, and air dry.

RESIDUE

Residue refers to material left in a vessel after evaporation of a water sample. It may be subdivided into organic matter and inorganic material. These may both be further subdivided into particulate and dissolved matter. Since variables are numerous in this type of determination, the accuracy is usually not as great as for most limnological determinations. Three procedures are described below. Total residue (sometimes referred to as suspended matter) is simply the weight of material remaining after evaporation of raw water. Filterable residue (dissolved matter or dissolved solids) is the weight of material remaining after evaporation of the filtrate of the raw water sample. Volatile matter (conversely fixed residue) is the difference between the weight obtained in either of the above procedures (specified) and that obtained after ignition at 600° C.

Total residue

Equipment

Evaporating dishes
Drying oven at 103° C
Analytical balance

Procedure

Collect sample in Pyrex or polyethylene bottle.

1. The dish to be used to hold the sample must be pretreated in the same manner as used on the sample (Fig. 13). This means that if the sample is to be dried in an oven and then placed in a desiccator before weighing, the empty dish must be taken through the same series of steps to determine the weight of the empty dish. If igni-

tion is to be used, the empty dish must be ignited, cooled in a desiccator, and weighed for initial weight.

2. To the prepared and preweighed dish add a well-mixed aliquot of water sample to be evaporated. Usually 100 ml will suffice, but exact volume will be determined by the sensitivity of the balance. It must not contain more than 200 mg of residue.

3. Evaporate to dryness, and then dry overnight in an oven at 103° C.

4. Remove dish from oven, cool briefly in the air, and place in desiccator; dishes must not touch each other or the sides of the desiccator.

5. Take desiccator to balance, remove cooled dishes, and weigh rapidly.

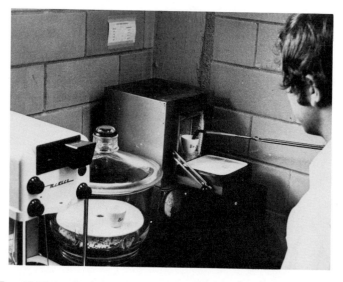

Fig. 13. The volatile matter portion of total residue is determined by igniting the evaporated and weighed sample in a muffle furnace at 600° C. The crucible containing the ash is cooled in a desiccator and reweighed. The difference in weights is volatile matter.

Calculations

mg total residue on drying at $103°$ C $= \dfrac{\text{mg total residue} \times 1{,}000}{\text{ml of sample}}$

Filterable residue (dissolved solids)

Equipment

As for total residue, plus a filter apparatus using acid-washed, ashless, hard-finish filter paper for fine precipitates, or for most precise work a filter apparatus using 0.45μm membrane filters

Procedure

The procedure is the same as for total residue except that the sample is filtered and the filtrate evaporated. Results are reported as filterable residue or total dissolved solids on drying at $103°$ C.

Note: In waters where soluble organic matter is not great, the total filterable residue is due primarily to mineral content of the water. Thus it may be used to check for completeness of analysis of the water for mineral content. A good correlation between the sum of the ion determinations ($Ca + Mg + HCO_3$ and so forth) and dissolved matter is conclusive that no major ion has been omitted from the analysis.

Since, as pointed out in the section on specific conductance, the concentration of dissolved ionic material may be generally related to the conductivity of the water, it is possible to estimate total dissolved matter from specific conductance data. This is accomplished by multiplying the specific conductance by some empirical factor that is usually less than 1. Rainwater and Thatcher (1960) suggest that 0.65 times specific conductance approximates filterable residue (total dissolved matter or solids) in most natural waters. However, it is best to determine this factor for each body of water routinely studied. This is done simply by determining the specific conductance at a stated temperature of the same water to be used for determination of filterable residue. The factor *(f)* is then determined by dividing the filterable residue by specific conductance. Once this is determined, one may avoid using the evaporation method on a regular schedule and employ it only to occasionally check the validity of the f factor.

Volatile matter

Volatile matter provides only a crude estimate of organic matter present in the water sample mainly because ignition also produces a decomposition or volatilization of inorganic salts. Better estimates of organic matter, especially in low concentrations, may be obtained by quantitative dichromate oxidation (Maciolek, 1962) or infrared carbon analysis. However, the volatile matter procedure is more simple because the sample used in determination of total or filterable residue is carried on through only one more step.

Equipment

As for residue, but vessel must be able to withstand higher temperatures and must be preignited before initial weight is determined; also, an electric muffle furnace is required

Procedure

1. Preheat muffle furnace to 600° C.
2. Ignite sample used in residue determination for 15 minutes, air cool briefly, complete cooling in desiccator, and weigh. The loss on ignition is reported as milligrams of volatile solids per liter.

Calculations

$$\text{mg volatile solids per liter} = \frac{\text{mg volatile solids} \times 1,000}{\text{ml of sample}}$$

THE
CHAPTER THREE   # PLANKTON

The plankton are those organisms that, because of their size or immobility or both, are at the mercy of water movements. The limnologist generally considers these to be tiny forms of life. However, this group encompasses organisms whose size spans three orders of magnitude. Plankton include forms of aquatic bacteria and ultra-algae only a few microns in diameter and macroscopic forms of crustacea several millimeters long. Thus the plankton encompass a wider range of sizes than do the larger, macroscopic nekton.

Most plankton sampling techniques are size selective. A convenient, if artifical, size differentiation of the plankton is net plankton and nannoplankton. The distinction is based on the passage of organisms through a no. 25 mesh plankton net (Table 6). The organisms retained by the net are considered net plankton, regardless of their taxonomic position. Although by no means absolute or precise, this distinction tends to separate the zooplankton (animal) from the phytoplankton (plant) species.

Plankton are important components of aquatic systems. Because they lack both motility and attachment devices, they are not commonly found in rivers. In larger lakes and reservoirs the plankton are the major primary and secondary links in the trophic relationship. This

TABLE 6. Plankton net bolting cloth specifications

Silk size no.	Meshes per inch	Silk aperture (μ)	Nylon aperture (μ)
0000	16	1364	1340
000	23	1024	1050
00	29	752	752
0	38	569	571
1	48	417	423
2	54	366	363
3	58	333	333
4	62	318	316
5	66	282	280
6	74	239	243
7	82	224	223
8	86	203	202
9	97	168	165
10	109	158	153
11	116	145	—
12	125	119	116
13	129	112	110
14	139	99	102
15	150	94	93
16	157	86	86
17	163	81	—
18	166	79	—
19	169	77	—
20	173	76	80
21	178	69	—
25	200	64	64

significance decreases as the volume of water decreases; in decreased volumes of water the benthic forms become increasingly important. Plankton organisms are also of interest to the biologist because of their physical nature. The ability to maintain position in an open water system requires that most of these organisms have unusual modifications for flotation. Many groups, although lacking obvious means of motility, do have regular diel vertical movements. The mechanisms and stimuli for such movements have interested aquatic biologists for some time. Also, the plankton often serve as indicators of water quality. Chemical conditions are known to determine the taxonomic nature of the plankton, and the species composition is often used not only to classify the water as polluted or free from pollution but also to determine the quantities of various naturally occurring substances, such as nitrogen and phosphorus.

In sampling the plankton, one must first determine what group of organisms are desired and from what portion of the aquatic system they are to come. Several devices for collecting these organisms have been devised. Three "net" methods and one "nanno" method are described below. A net towed behind a boat for some distance provides the greatest quantity of plankton with minimum effort. The most advanced design of this type is the Clarke-Bumpus net, which is a net with a metering device to determine the quantity of water passing through it. Although the net is excellent in many respects, its major weakness is that precise depths are difficult to sample because a slight change in the movement of the boat causes the depth of the net to vary.

A second method employs the Wisconsin net (this net can likewise be used as a tow net), in which known quantities of water are collected from discrete depths by the use of a water sampler or pump and are poured through the net. This method assures collection from precise depths but has a serious drawback in that the size of the sample is relatively small because of the time and effort involved in collecting many samples with the usual 1- to 3-liter water-sampling bottles. A plankton pump that empties through the net overcomes the volume problem but introduces additional mechanical problems in the field.

A third method employs the Juday trap, which provides precise sampling of a given depth but also has the severe problem of small sample size. This trap is a metal box that is lowered to the desired depth with the top and bottom open. The openings are tripped by messenger to snap shut and encompass a cubic volume of water, which drains through a plankton net. The principle advantage of this sampler is the ability to sample precise depths. However, in addition to the problem of small sample size, a second disadvantage is the delicate nature of the device, especially the sliding-door mechanisms. Nevertheless, the fact that a sample is both collected and concentrated in one step makes this type of sampler a favorite of many investigators of the net plankton.

Nannoplankton are those organisms that pass through a No. 25 net. To collect these organisms, it is obvious that water must also be passed through such a net and be collected. The tiny organisms must later be separated from the water by some other method, such as settling, filtration, or centrifugation, all of which are done in the laboratory. The usual field technique is to collect the water that was passed through the Wisconsin net as described above and preserve it for further nannoplankton analysis.

The following procedures describe methods of collecting plankton from discrete depths. Although collection from separate depths is usually the preferred technique for quantitative studies, it is possible to entirely miss some plankton organisms because of their presence in a thin "plate" located between the preselected sampling depths. A vertical tow started at a selected depth or at the bottom will sample plankton at all depths. Although vertical tows are primarily used for qualitative work, quantitative data may be obtained by calculating the volume of water sampled as follows: volume sampled = length of tow $\times \pi \times$ (radius of net opening)^2. If the length of vertical tow is insufficient to give an adequate concentration of organisms, an oblique tow should be made, from a preselected depth or from the bottom, behind a moving boat. It is important that the net be retrieved to the surface slowly and at a regular rate to avoid bias of oversampling or undersampling at any one depth.

CLARKE-BUMPUS NET

Apparatus

Clarke-Bumpus net with 2 messengers, buckets, and calibrated line

4- to 6-oz. bottles or baby food jars

FIG. 14. Clarke-Bumpus metering plankton net. This photograph shows the net with the door in the intermediate, or open, position. When the net is being towed, the lateral planes cause the net to ride in a position approximately at a right angle to the metal suspending frame.

Plastic "squeeze" bottle
Clinometer

Reagents

Supply of neutral formalin (saturate 37% formaldehyde
solution with magnesium carbonate)

Procedure

1. Select portion of lake to be sampled, assemble
Clarke-Bumpus net, using net of desired mesh for the type
of organisms wanted. Be sure that net and bucket mesh
are of the same size. Attach free end of calibrated line to
boat.

2. Close door mechanism of net, take and record meter
readings, and lower net over the side of the boat. The
depth to be sampled may be determined by the use of the
trigonometric function for a right triangle. The line trailing
behind the boat forms the hypotenuse of this triangle.
Using a clinometer (one can easily be made from a pro-
tractor), determine depth of tow by multiplying length of
hypotenuse by the cosine of the angle as measured by the
clinometer, or read from nomograph (Fig. 15).

3. Once the boat speed is maintained and steadied and
the desired depth is reached, drop the first messenger
down the line to open the net. Tow for a measured period
of time (try 1 minute to start), and then drop the second
messenger to close the net.

4. Retrieve net and wash up and down while lifting the
net from the water to gradually wash organisms from the
sides of the net down into the bucket. Take meter reading.
The number of revolutions turned while the net is opened
is proportional to the volume of water passing through the
net. This relationship of volume to revolutions is usually
determined by the manufacturer and recorded for that
particular net; however, with time, corrosion, and wear,
this calibration changes, and it is important to recalibrate
the net's meter by moving it through a known distance in
the water and calculating the volume passed through per

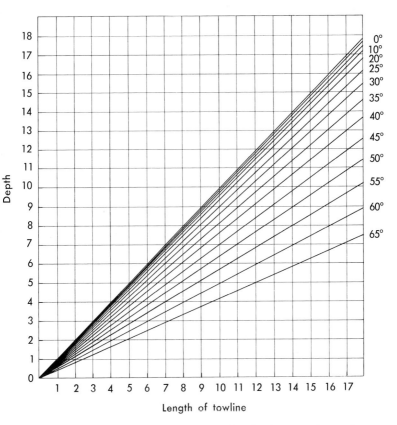

FIG. 15. Nomograph for determining depth of a towed sampler when length of towline and angle of tow are known.

revolution. Since you recorded the time of the sampling run for a given boat speed, you now have an approximation of how long a net must be in the water at that speed to filter a desired quantity of water. This is useful for future tows.

5. Disassemble the bucket, and wash the organisms into one of the jars. Use a stream of water from the plastic squeeze bottle to wash the last of the organisms into the

jar, and then preserve with formalin to produce a 4% to 5% solution. Label the bottle with data for the collection.

6. Wash out net thoroughly and reassemble for next tow. It is important that all plankton nets always be well washed at the end of a day's sampling and preferably again when the nets are returned to the laboratory. This prevents gradual plugging of the nets and possible carry-over of organisms from one sample to another.

WISCONSIN NET FOR NET PLANKTON AND NANNOPLANKTON

Apparatus

No. 20 or 25 Wisconsin plankton net
Plastic pail
Kemmerer or van Dorn water-sampling bottle
4- to 6-oz. bottles or baby food jars
1- to 2-liter polyethylene bottles

Reagents

Supply of neutral formalin or Lugol's iodine (5 g I, 10 g KI, 10 ml glacial acetic acid in 100 ml distilled water)

Procedure

1. Select a sampling site, and anchor boat to prevent drift. Choose sampling depths based on physical, chemical, or morphological features of the lake.

2. Assemble Wisconsin net with bucket, and collect water samples from the desired depth. These samples are strained through the mouth of the Wisconsin net, and the water is collected in the plastic bucket. Be sure no water splashes over the outside edge of the Wisconsin net. Count and record (it is easy to forget) number of water samplers emptied through the net. Determine volume of water filtered by multiplying the number of samplers emptied by the volume of the sampler used. Although the actual volume necessary to provide a concentrate adequate for most studies will vary widely depending on the nature of the water, a minimum of 30

FIG. 16. The popular Wisconsin plankton net. It may be used as a tow net, either vertically, obliquely, or horizontally, or as a filter net through which water is poured. (Courtesy Wildlife Supply Company, Saginaw, Michigan 48602.)

liters is usually required in eutrophic waters and much more in oligotrophic waters.

3. Wash down organisms inside plankton net by sloshing it over side of boat, but not allowing water to flow over mouth of the net, while gradually raising net from the water so that organisms are concentrated in the bucket. Disassemble bucket, and while holding it over the mouth of the jar, remove stopper and allow contents to drain into jar. Wash down sides of this jar with enough formalin to make a 4% to 5% solution, and label with collection data.

4. Pour the material that passed through the net but was collected in the plastic bucket into the large plastic bottle, preserve with adequate formalin to make a 4% to 5% solution, and label with collection data. Many investigators prefer other preservatives for nannoplankton. One of the most commonly used is Lugol's iodine. Sufficient Lugol's iodine is added to the water to make a 1% solution (color of weak tea).

JUDAY TRAP

Apparatus

Juday trap, messenger, plankton bucket of appropriate mesh, and calibrated line

4- to 6-oz. bottles or baby food jars

Plastic squeeze bottle

Reagents

Supply of neutral formalin (saturate 37% formaldehyde solution with magnesium carbonate)

Procedure

1. Attach line to trap by passing it through hole in trip lever and then securing it beneath hole in plate attached to the four suspending chains (if a solid messenger is used, slip on line before attaching trap). Inspect grooves of sliding door mechanism for cleanliness, and

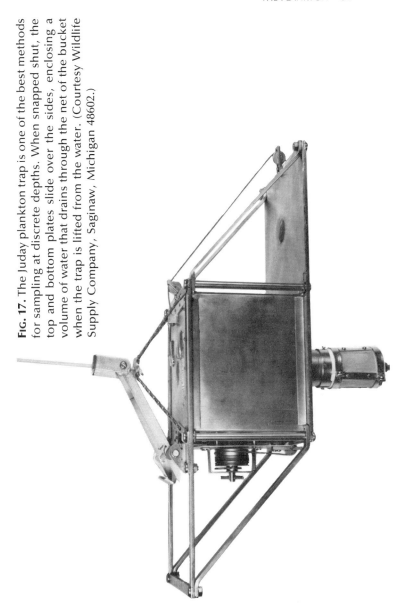

Fig. 17. The Juday plankton trap is one of the best methods for sampling at discrete depths. When snapped shut, the top and bottom plates slide over the sides, enclosing a volume of water that drains through the net of the bucket when the trap is lifted from the water. (Courtesy Wildlife Supply Company, Saginaw, Michigan 48602.)

check freedom of movement of doors. Attach plankton bucket. Open doors and set trip lever mechanism's knife edge behind block (weight of trap must be kept on line from this point).

2. Lower trap to desired depth. Small traps up to 5-liter capacity may be lowered by hand, but larger traps require a winch and boom. As the trap is lowered, a column of water is cut so that at any instant the contained volume represents those organisms from that depth.

3. Drop messenger to close trap. Although avoidance by organisms is minimal as the trap descends, larger and more motile forms may escape from the trap if there is unnecessary delay in closing. With practice, one can drop the messenger slightly before trap reaches desired depth so that closure and time of reaching that depth are almost simultaneous.

4. Retrieve trap and remove bucket. Drain off excess water, and wash organisms into jar with a stream of water (filtered lake water is all right) from squeeze bottle. Add sufficient neutral formalin to produce a 4% to 5% solution. (Some prefer to use neutral 70% isopropyl or ethyl alcohol as a plankton preservative. When this is done, use 70% alcohol in the squeeze bottle to wash plankton into the jar.)

5. For most water, a single 5-liter sample is inadequate to assure representation of most species or to assure quantitative sampling accuracy. Consequently, replicates are taken from the same site and depth. These may be pooled in a single jar for survey work or kept separate where statistical analysis is desired. Twenty liters is often an adequate volume for many mesotrophic to eutrophic lakes and reservoirs.

6. Before storing trap, inspect grooves for cleanliness. It is also good practice to spray the slides, grooves, and springs with an aerosol water-dispersing lubricant, such as WD-40.

NANNOPLANKTON LABORATORY METHODS
Nannoplankton by membrane filter

Any method for the identification and enumeration of nannoplankton must permit examination of these materials under a compound microscope, using a minimum of the high-power dry objective ($\approx$430$\times$) and preferably using oil immersion ($\approx$970$\times$). Many of these organisms are delicate, and preparation for examination requires attention to avoid fracturing, rupturing, or plasmolyzing them.

Apparatus

1-inch diameter gridded membrane filters, pore size
   0.8μm
Vacuum filtration apparatus
Graduated cylinder
Clearing oil (immersion or cedarwood oil)

Procedure

1. Measure out a portion of the well-shaken, preserved nannoplankton sample into a graduated cylinder. In many waters 25 ml may be appropriate.

2. Filter this quantity through the membrane filter, using very low vacuum. Add contents of graduated cylinder all at once to promote a random distribution of organisms over the surface of the membrane filter.

3. Remove filter from apparatus, handling by the edges, and place filter on cleaned microscope slide. Align horizontal lines of grid parallel to edge of microscope slide.

4. Place a drop of clearing oil on the microscope slide at each edge of membrane filter. *Do not* cover filter with oil. The center portion of the filter must remain oil-free to permit evaporation of water for drying of the slide. As the slide dries, the oil from the edges will replace the water (Fig. 18) from the edges. When dried, the membrane filter will be transparent. A milky filter

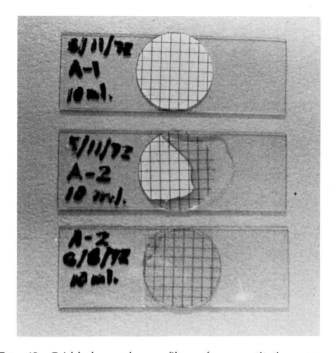

FIG. 18. Gridded membrane filters for quantitative nanno-plankton analysis. An aliquot of lake water is gently filtered and the damp filter immediately placed on a clean microscope slide *(top)*. Several drops of oil are placed on the slide at the edge of the filter. As the filter dries, oil is drawn into the filter matrix, rendering it transparent *(center)*. The completely oil-impregnated clear filter is now ready for microscopic examination *(bottom)*.

coloration indicates that the filter became oil-coated without complete drying. The time for drying and clearing depends on the air moisture conditions but is usually complete within 2 to 4 hours. During this drying, the slide should be stored in a dust-free place.

5. The microscope slide is now ready for examination with or without a cover slip. One may work with

the oil immersion lens directly on the membrane filter if an additional drop of oil is added. With this preparation, most organisms may be both identified and counted on the same slide. Common forms of life observed in the nannoplankton will be phytoplankton, rotifers, and some protozoa.

6. Organisms may be counted by the total count, strip count, or field count method, depending on the numbers of organisms present. A simple method to estimate when sufficient fields have been counted is the changing average method. To do this, count from an initial selected set of fields (for example, 10 fields). Analyze these counts and select an organism that appears regularly but not in high numbers. Calculate the average number of these organisms per field. Then count an additional field, add the number of organisms, and recalculate the average. If the average was not changed by the additional field, then sufficient fields have been counted; however, if the average was changed by the additional field, continue to count additional fields until no further change in the average frequency of that organism is determined. A tabulation sheet such as that shown in Fig. 19 is helpful.

Calculations

Number of organisms of each species per aliquot filtered =

$$\frac{\text{number of organisms in fields counted}}{\text{fields counted}} \times \text{number of fields in filtered area}$$

Number of organisms of each species per liter of lake water =

$$\frac{(\text{number of organisms in aliquot filtered}) \times 1,000}{\text{volume filtered (ml)}}$$

PREPARATION OF PERMANENT MEMBRANE-FILTERED NANNOPLANKTON SLIDES

The membrane filter method using immersion or cedarwood oil provides relatively permanent slides if kept in a dust-free place. Cedarwood oil tends to evapo-

PLANKTON ENUMERATION

Water_____ Station _____ Date _____

Hour Collected_____Wind_____ Sky_____

Depth_____Net Size _____Volume filtered _____Collected by _____

Volume Concentrate_____Counting Method _____Count by _____

Organisms	Field Strip counts										Total	Average number per ml concentrate	Number per liter of original H_2O
	1	2	3	4	5	6	7	8	9	10			
Totals													

Remarks:

FIG. 19. Use of a standard form for enumeration of plankton is convenient and assures that all necessary data are permanently recorded. Identified organisms are listed in the lefthand column. Space is provided to record the number present in each of ten different field or strip counts.

rate more rapidly than immersion oil and must be replaced occasionally, but both will provide slides that will last for months. If permanent slides are desired, the following procedure may be used.

Reagents

Glutaraldehyde-ethanol solution. Add five drops of glutaraldehyde, with mixing, to 50 ml distilled water. Refrigerate for 6 hours. Then mix with 50 ml ethyl alcohol.

Beechwood creosote–ethanol mixture. Mix 7 ml of 95% ethyl alcohol with 3 ml beechwood creosote.

Procedure

1. Gently filter sample as above, but do not filter dry. Leave approximately 5 ml water in the filter apparatus, and add 10 to 15 ml of the glutaraldehyde-ethanol mixture without mixing. Filter down to 5 ml.

2. Add 10 ml of the beechwood creosote–ethanol mixture, and filter until only a very thin liquid film remains on the filter.

3. Remove filter and place on microscope slide. Saturate with pure beechwood creosote. Add cover slip, and set aside for approximately 2 weeks to clear.

4. Clean filtration apparatus with 95% ethanol.

Nannoplankton by sedimentation tube–inverted microscope

The sedimentation tube–inverted microscope method has the distinct advantage of being the most gentle for soft-bodied forms. It provides the most transparent slides and, therefore, is best for identification requiring work with minute structures. The major disadvantages are the time required for the sedimentation process to occur and the necessity of an inverted microscope. For all but advanced studies, it is doubtful that the advantages of this method would offset the disadvantages when compared with the membrane filter technique.

Apparatus

Inverted microscope.

Sedimentation cylinders, which may be purchased or made in a variety of sizes. Typical size ranges are from 5- to 25-ml capacity. To construct these in a laboratory, one may use either glass or Lucite tubing 22 to 25 mm in diameter. A microscope slide or a large rectangular or circular glass cover slip is cemented to one end with a minimum of cement. To assure complete sedimentation of small forms, the height of the contained water column must not exceed five times the diameter of the cylinder. A small, flat square of Lucite or glass should also be available to serve as a temporary cover for the cylinder.

Procedure

1. Measure out a known volume of a well-shaken nannoplankton sample that has been preserved in Lugol's iodine, and pour into the cylinder.
2. Cover and set cylinder aside in a vibration-free place near the inverted microscope to allow sedimentation to occur. The time required for sedimentation of small organisms can be calculated by allowing 4 hours for each centimeter of water column.
3. Following sedimentation, place cylinder on an inverted microscope stage, and take either total, field, or strip counts. The magnification under which these counts are made will vary with the size of the organism encountered. Make a cursory examination of the cylinder under oil immersion to determine if there are very small forms present. If so, identification and enumeration should be at this magnification. A total count in even moderately rich waters under this magnification would likely require many hours. In this case, count random fields, using a Whipple ocular disk that has been calibrated by stage micrometer; or, using opposite parallel edges of the Whipple disk, count strips across the field until adequate numbers are obtained. Count until there is no further change in the average occurrence per field as indicated above or, as some investigators prefer, until you have counted 100 organisms of average frequency.

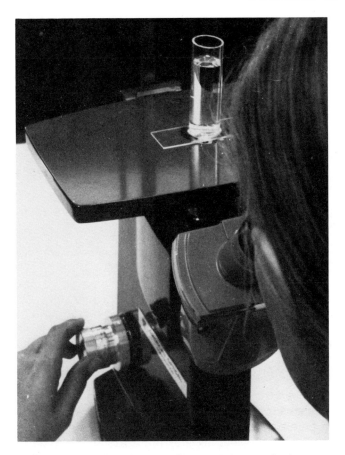

FIG. 20. Use of sedimentation tube on an inverted microscope is especially valuable in the analysis of delicate plankton organisms that may be unrecognizably damaged by other techniques of concentration.

NET PLANKTON

Net plankton may be identified and enumerated by either aliquot methods or total count methods. Total counts are usually taken of the larger forms, such as copepods and cladocerans, as well as colonial algae, such as *Volvox*. The aliquot methods are usually used for rotifers, colonial diatoms, such as *Asterionella* or *Tabellaria,* and filamentous green and blue-green algae. Copepod nauplii are also usually counted by aliquot methods. If specific identifications are to be made, it is often best to identify the individuals before beginning actual enumeration, since many forms require micromanipulation or dissection. The copepod and cladocera populations at any given time are usually restricted to three or four species of copepods and four to six species of cladocera; usually one or two of each are dominant.

Smaller forms by aliquot

Apparatus

Compound microscope with mechanical stage and equipped with Whipple eyepiece disk

Dissecting microscope

Sedgwick-Rafter counting cell with cover slip

Hensen-Stempel pipette or large-bore pipette with 1- to 2-ml capacity

Petri dish, 5- to 10-cm diameter

Procedure

1. Mix preserved sample well, and withdraw aliquot by means of Hensen-Stempel pipette or by quick suction on regular pipette. Introduce this sample into Sedgwick-Rafter counting cell as illustrated in Fig. 21. When the cover slip is placed diagonally and the sample is introduced in one or both corners, a surface film of water raises the cover slip from the ground glass edges and allows it to swing about to cover the cell. Be sure no bubbles form in the corners. To avoid formation of bubbles during counting period because of warming of sample, be sure the sample is at or slightly above room temperature before making the slide.

2. Place slide on stage of microscope and take either total, strip, or field counts as necessary to obtain required number of organisms for statistical validity, count-

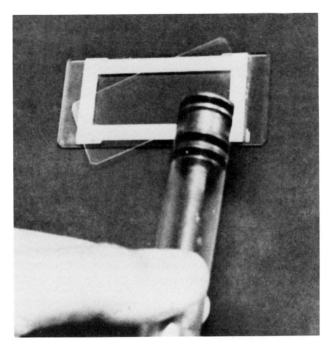

Fig. 21. The Sedgwick-Rafter cell is commonly used for quantitative analysis of net plankton. One ml of sample is introduced at the corner of the cell by means of a Hensen-Stempel pipette. This pipette is designed to take a random aliquot from the plankton sample.

ing until average does not change or until 100 organisms of average frequency are counted. It is quicker to count these organisms using a 10× objective with 10× oculars; however, many investigators prefer to use a 20× objective, which improves the opportunity for identification on the Sedgwick-Rafter cell. The 20× objective is the maximum that the depth of the Sedgwick-Rafter cell will permit.

3. For many waters, three longitudinal strip counts at 100× magnification will be adequate. The upper and lower parallel sides of the Whipple eyepiece disk are used

to limit the width of the counted strip. Organisms falling on the upper line are included in the count, whereas organisms falling on the lower line are excluded. When $10\times$ objectives are used, the width of the strip counted with a Whipple disk is approximately 1 mm; however, the width should be accurately measured with a stage micrometer. If we assume a width of 1 mm, a strip length of 50 mm, and cell depth of 1 mm, each strip will have counted 50 mm^3, or 1/20 of the volume of the cell. From these relationships, calculate the total number of each species in the 1 ml of water contained by the Sedgwick-Rafter cell.

4. These counts are for 1 ml of plankton concentrate. It is now necessary to relate these data back to the content of unconcentrated lake water. To do this, measure the volume of concentrate. Then determine the number of organisms of each species per liter of lake water as follows:

$$\text{Organisms per liter of lake water} = \frac{\text{organisms per ml of concentrate} \times 1,000}{\text{concentration factor}}$$

where

$$\text{Concentration factor} = \frac{\text{volume of lake water filtered (ml)}}{\text{volume of concentrate (ml)}}$$

Larger forms by total count
Procedure

1. Empty contents of concentrated preserved sample from bottle into Petri dish with grid or strips on the bottom. If the volume is too great to be contained in this dish, it may be further concentrated in the laboratory by pouring through a portion of plankton netting.

2. Allow this material to settle for a few minutes, and then count by either total, field, or strip method. For larger forms, such as copepods and cladocera, use a magnification of $20\times$ or $30\times$; consequently, it is easy to count the entire sample for these large forms.

3. Obtain the number of organisms of each species per original liter of lake sample by determining the volume of concentrate in the dish and dividing this into the total volume filtered (be sure to use the same units) to determine a concentration factor.

PLANKTON CHLOROPHYLL *A*

The following is a general survey method for estimating the concentration of phytoplankton chlorophyll *a* pigment in a volume of water. The value of such data is uncertain, since many variables are present. However, among the many attempts to use such data are characterization of community age and structure (Odum, McConnell, and Abbott, 1958), quantification of phytoplankton standing crop (Small, 1961), and photosynthetic rates (Ryther and Yentsch, 1957).

Apparatus

Spectronic 20 with red phototube and filter
1-inch matched test tubes
1-inch regular test tubes
Rubber stoppers or corks to fit 1-inch regular test tubes (must be soaked in dilute acetone for 24 hours to leach out any color)
Membrane filter unit with $0.8\mu m$, 47-mm filters
Centrifuge with stoppered *glass* centrifuge tubes

Reagents

90% alkalized acetone. Using reagent grade acetone and distilled water, prepare a 90% solution. Add about one spoonful of powdered magnesium carbonate per liter of acetone solution.

Procedure

1. Measure depth of photic zone with submarine light meter between 10:00 AM and 3:00 PM when sun is not occluded by clouds.
2. Take water samples at several random loca-

tions over lake and at several depths down to limit of photic zone. Usually 1-liter samples will be adequate.

3. Filter portion of well-shaken sample through $0.8\mu m$, 47-mm membrane filter. Because membrane filters become clogged suddenly, in contrast to the gradual decrease in filtering rate that occurs with paper filters, pour 25- to 50-ml portions of lake water into the receiving funnel, waiting for each portion to be filtered before adding more water. When filtering rate begins to decrease perceptibly, the filter is about to clog completely. At this point allow suction to damp-dry the filter, and then *carefully* remove receiving funnel from sintered glass filter support. Do not decrease suction until edge of filter has been lifted to vent suction flask. At this time suction may be turned off and the filter carefully removed. Do not touch plankton on filter; hold by edges only.

4. Roll up or fold filter with plankton inside. Place filter in bottom of dry 1-inch test tube.

5. Measure volume of water filtered. Save remainder of unfiltered sample.

6. Add 20 ml of 90% alkalized acetone to sample tube containing filter and plankton. Stopper tube with an acetone-leached stopper. Shake until filter is dissolved. If acetone is not a distinct green or olive color, repeat steps 3, 4, and 5, and place additional filter in the tube with the first one until solution is visibly green. Solubility of membrane filters is about one 47-mm filter per 5 ml acetone. Therefore, four filters is the maximum usable number.

7. Allow extraction to proceed in a dark refrigerator. An overnight extraction is preferable.

8. If any turbidity is apparent in extract, centrifuge extract in stoppered tubes.

9. Use a blank of 20 ml of acetone containing the same number of filters used for sample, and measure absorbance of sample and blank at 663 nm. In acetone ex-

tracts the absorbance of this turbidity blank may be appreciable.

Calculations

For grass green extracts:

$$\frac{\text{Chlorophyll } a \text{ concentration}}{\text{(mg per liter of extract)}} = \text{absorbance} \times 7.5^*$$

Chlorophyll a is the chief chlorophyll measured at 663 nm; however, the band width of the B & L Spectronic 20 probably also measures some chlorophyll b when it is present.

Chlorophyll a concentration of original lake water (mg/m^3) =

$$\frac{\text{chlorophyll in extract (mg/liter)} \times \text{extract volume (ml)}}{\text{filtered volume (liters)}}$$

If a narrow band pass (<5 nm) spectrophotometer is available, the equations from Strickland and Parsons' handbook (1968) are preferable. These not only improve accuracy of chlorophyll a determination but also allow estimations of chlorophylls b and c.

Procedure and calculations for use with a narrow band pass spectrophotometer

Prepare 20 ml extract as above, but measure absorbance (abs.) at 750, 665, 645, and 630 nm. Subtract absorbance at 750 nm from each of the other absorbances to correct for turbidity. The following calculations are given for a spectrophotometer cell of 1-cm light path.

Chlorophyll a concentration (C_a) in sample = 116 (abs. at 665 nm) − 13.1 (abs. at 645 nm) − 1.4 (abs. at 630 nm)

Chlorophyll b concentration (C_b) in sample = 207 (abs. at 645 nm) − 43.3 (abs. at 665 nm) − 44.2 (abs. at 630 nm)

Chlorophyll c concentration (C_c) in sample = 550 (abs. at 630 nm) − 46.4 (abs. at 665 nm) − 163 (abs. at 645 nm)

$$\text{mg pigment per m}^3 \text{ of lake water} = \frac{C \times 2}{\text{liters of lake water filtered}}$$

*7.5 is used for 1-inch cuvettes; 13.4 is used for 1-cm cuvettes.

PHYTOPLANKTON PHOTOSYNTHESIS BY OXYGEN METHOD

Many different parameters can be used to determine the synthetic rate of phytoplankton. One can make direct measurements of increases in carbon, dry organic matter, plankton biomass, or dry plankton. However, the most common method (and the quickest) is the measurement of oxygen evolved during photosynthesis. Since we know the approximate relationship of oxygen evolution to carbon reduction and carbohydrate formation, photosynthesis can be stated in various terms, equivalent to the other parameters. Table 7 indicates phytoplankton equivalents and permits the conversion of oxygen evolution data to other terms (or, indeed, interconversion between any of the various equivalent parameters of production). For example, the evolution of 1 ml of oxygen (column 3 of Table 7) is equivalent to the synthesis of 0.53 mg carbon, 1.0 mg dry organic matter (1.0 ml oxygen), 1.44 mg oxygen, 4.7 mg plankton biomass, and 1.05 mg dry plankton.

The light- and dark-bottle method (Gaarder and Gran, 1927) of measuring the synthetic rate of phytoplankton

TABLE 7. Freshwater phytoplankton equivalents

	Carbon 1 mg	Dry organic matter 1 mg	Oxygen 1 ml	Oxygen 1 mg	Plankton biomass 1 mg	Dry plankton 1 mg
Carbon (mg)	1.0	0.54	0.53	0.375	0.11	0.50
Dry organic matter (mg)	1.85	1.0	1.0	0.69	0.20	0.92
Oxygen (ml)	1.87	1.0	1.0	0.70	0.21	0.93
Oxygen (mg)	2.67	1.44	1.44	1.0	0.29	1.34
Plankton biomass (mg)	8.9	4.8	4.7	3.3	1.0	4.45
Dry plankton (mg)	1.98	1.07	1.05	0.74	0.22	1.0

is suitable for eutrophic waters. Precise analysis of small changes in oxygen concentration may be difficult in oligotrophic waters. This method compares the oxygen changes that occur in plankton communities contained in clear bottles with those occurring in dark bottles during a 24-hour period. In the light bottles, oxygen is evolved during photosynthesis and consumed by plant and animal respiration. In the dark bottle only respiration occurs.

Apparatus

Three glass-stoppered 125- to 300-ml reagent or BOD bottles for each measurement, one painted black (or wrapped with black plastic tape) and wrapped in aluminum foil (provide a square of aluminum foil for final capping)

Rope or other means to suspend pairs of bottles in lake at depth of collection

Various means of attaching paired bottles to rope (snaps and rings, clamps, or wire baskets)

Water collection and analysis apparatus as described under dissolved oxygen

Reagents

As for dissolved oxygen

Procedure

1. Select stations that will adequately represent the major morphological characteristics of the lake and depths that will adequately represent conditions of the water column. These depths may be at regular intervals or at percentages of surface illumination (100%, 75%, 50%, 25%, 1%).

2. Take a sample from each depth, using the best oxygen techniques. Use two light bottles and one dark bottle for each depth. It is preferable to use a large enough sampler (3 liters) to fill all three bottles.

3. Return one light and one dark bottle to their original depth for incubation. Be certain that glass

stoppers are firmly in place and that foil cap completely seals area between cap and shoulder of the dark bottle. It is important to avoid even pinhole light leaks. Record the time.

4. Fix remaining bottle from each depth with Winkler oxygen reagents, and determine initial (I) oxygen concentration of the water sample by titration.

5. After 24 hours recover other bottles and determine oxygen concentration.

Calculations

Plankton community respiration (as mg oxygen consumed per hour $= \dfrac{I - D}{24}$

Net plankton photosynthesis (as mg oxygen per hour) $= \dfrac{L - I}{24}$

Gross plankton photosynthesis (as mg oxygen per hour) = net photosynthesis + community respiration

where

I, D, and L = oxygen concentrations as mg per liter for initial bottle, dark bottle, and light bottle, respectively

These data may be converted to grams of oxygen per cubic meter by multiplying numerator and denominator by 1,000. To estimate the photosynthetic activity per square meter of lake surface, either (1) multiply cubic meter value by depth, in meters, that it represents, and sum each of these; or (2) plot actual cubic meter values on linear graph paper against depth (m), and then integrate area beneath curve by planimeter or counting of squares.

PHYTOPLANKTON PHOTOSYNTHESIS BY ^{14}C METHOD

The principal advantage of measuring phytoplankton photosynthesis by the ^{14}C method rather than using the oxygen change procedures is its greater sensitivity. This permits estimation of primary production in more oligotrophic waters. Its disadvantages include (1) higher costs of instruments and supplies, (2) the need to obtain the proper license when using nonexempt quantities of the isotope, and (3) some uncertainty about the exact nature of the rates being measured. It is assumed that when short incubation periods (1 to 4 hours) are used, the rate approximates net photosynthesis. This method is also widely used to measure photosynthesis in algal bioassay studies.

Apparatus for inoculation and incubation

Incubation bottles, 125- to 300-ml glass-stoppered bottles: an equal number of dark bottles are necessary and prepared as described under the oxygen method.

Support for incubation bottles: as under oxygen method.

Water collection bottle: inert polyvinyl chloride (PVC) type preferred but not essential for fresh waters.

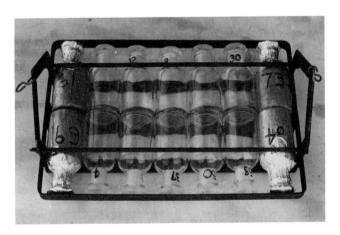

FIG. 22. When several replicates from one depth are to be incubated for ^{14}C uptake, some type of rack to resuspend all samples horizontally at one depth with a minimum of shading is desirable. Note the foil seal over the cap of each dark bottle.

Total carbon bottles: BOD or similar bottles for subsequent determination of total alkalinity and pH for each depth sampled.

Insulated "ice" chest painted black inside.

Aluminum foil or absorbent paper to contain radioactive spills.

Inoculation syringe with 4-inch cannula: a 1-ml tuberculin syringe with Chaney adaptor provides good replication.

Sodium carbonate–^{14}C solution, specific activity = 2 microcuries/ml: for occasional work one may purchase this as a sterile aqueous solution in sealed ampules with sufficient volume for the desired study. For extensive studies, it is much more economical to purchase concentrated solutions and dilute, then fill and seal ampules in your laboratory as follows: purchase $Na_2{}^{14}CO_3$ stock solution with a radioactive concentration of 2.0 millicuries/ml. Dilute 1 ml of stock solution to 1,000 ml with a dilution solution (dissolve 50 g NaCl in 1 liter of distilled water; add 0.3 g Na_2CO_3 and one pellet of NaOH). Fill ampules with quantity desired and seal. Autoclave ampules in inverted position in a metal pan filled with a diluted methylene blue solution. Remove from autoclave, and cool to room temperature in pan of dye solution. Any ampules that have an imperfect seal will suck the dye solution inside. These must be discarded. The absolute activity of the final solution will depend on the degree of accuracy claimed by the manufacturer of the stock solution and on your quantitive techniques. For precise work, the assumed activity must be confirmed. Such a service is provided for a fee by many commercial laboratories, or the activity may be determined in the laboratory if liquid scintillation instruments are available (Wolfe and Schelske, 1967).

Supplies

Rubber or disposable gloves

Containers for radioactive waste solids and liquids

Squeeze bottle with 5% HCl for decontamination of
spills

Apparatus for filtration and counting

Membrane filtration apparatus and source of vacuum:
47-mm diameter if counting is by liquid scintillation, or 25-mm diameter if counting is by planchet

Membrane filters, 0.45μm pore size and of correct
diameter

Source of distilled water

Small forceps for handling filters

Aluminum planchets and moderately fast-drying
household cement (Duco) if planchet counting, or
liquid scintillation vials if using liquid scintillation
for counting

Wax pencil or felt-tipped pen for labeling planchets
or vial caps

Provision for disposal of waste liquids and solids

Liquid scintillation "cocktail" as discussed below

Apparatus for determination of pH and total alkalinity

Procedure

1. Select locations and depths as for oxygen method.
Since this procedure takes only a short time, use care in
selecting time of incubation. Because of the possibility
of respiration of the radioactive compounds and conse-
quent underestimation of primary production, use the
shortest time period possible that will give sufficient
activity for precision in counting. A 2-hour midday period
will often provide sufficient illumination. A diel series of
measurements should be made to determine the relation
of the selected incubation time to the full photic period.
Generally, one half of the full day's photosynthesis occurs
during the 4-hour period from 10:00 AM to 2:00 PM.

2. Collect water samples from desired depths. Fill one light bottle, one dark bottle, and one bottle for pH and alkalinity from the sampler. Proceed with inoculation immediately; or, if several depths are to be sampled, place bottles in blackened ice chest until all samples for the location are taken; then proceed with inoculation as soon as possible.

3. Open an ampule and fill syringe, being certain that no bubbles remain.

4. Open the light and dark bottles and inoculate the $Na_2^{14}CO_3$ solution at bottom of bottle, using the long cannula. Remove cannula quickly, and restopper bottle. Immediately cap black bottle with a double thickness of foil. Inspect for tears or other possible light leaks. Begin timing incubation period.

5. Resuspend bottles at desired depth as soon as possible. A rack to maintain bottles in a horizontal position is desirable (Fig. 21). At all times, avoid exposing clear bottles to direct sunlight at surface of lake or while inoculating. Such exposure, especially of samples from greater depths, may cause a photoinhibition of the contained phytoplankton.

6. During incubation period, one may determine pH, total alkalinity, and temperature of sample if portable equipment is used or if a laboratory is nearby. If not, place these samples on ice for analysis upon returning to laboratory.

7. At end of incubation period, retrieve bottles and place in blackened ice chest. Cover them with crushed ice if field filtration is not to be done. Since samples in the iced-dark condition will not have a significant change in radioactivity for up to 3 hours, it may be preferable to return the samples to a laboratory where better filtration and vacuum is available.

8. Filter all of sample, or an aliquot thereof, depending on the amount of radioactive uptake and the presence of silt or detritus that clogs the filter. Gentle filtration is important to avoid cell damage and loss of tagged carbon

compounds; one-third atmosphere is probably safe for most forms. If a vacuum gauge is not available, use thin-walled rubber tubing rather than the usual thick-walled vacuum tubing. Often 50 ml taken as a well-shaken aliquot will provide sufficient radioactivity for short but precise counts. Wash filter and apparatus with a volume of distilled water approximately equal to the original volume filtered. If deposition of insoluble carbonate salts on the filter is suspected (alkaline water of high pH), remove any radioactive inorganic deposits from the filter by rinsing with 2% HCl or fuming over HCl vapors. Follow this procedure with a distilled water rinse, especially if liquid scintillation counting is to be used, since acid filters will not clear adequately in the scintillation fluid.

9a. Planchet counting: with fingertip, smear a thin film of cement on planchet, and immediately place filter on cement. A few seconds' delay will allow a surface film to dry; and attachment of filter may be imperfect, with subsequent curling. Write appropriate label on bottom of planchet. Count a known standard (to determine efficiency of counter) and the samples according to instructions for detector system used. Count to a minimum of 2,000 (approximately 5% error). A small sample size, such as 50 ml, will usually produce a sufficiently thin layer so that self-absorption can be ignored.

9b. Liquid scintillation counting: if instrumentation is available, liquid scintillation counting offers many advantages (including a much greater efficiency) over planchet counting (Lind and Campbell, 1969). Remove filter and coil to place vertically in scintillation vial, with algal layer to the inside. Two scintillation cocktails are commonly used. One requires desiccation of filters before adding cocktail but gives higher efficiency. The second, though it has a slightly lower efficiency, does not require desiccation, which permits a more rapid completion of the procedure and recovery of data. If the former is desired, place uncapped vials in a desiccator charged with silica gel desiccant and small vial of CO_2 absorbent.

When dry, add 20 ml of the cocktail: 4 g of PPO (2,5-diphenyloxazole) and 0.1 g of dimethyl POPOP (1,4-*bis*-2-[5-phenyloxazolyl]-benzene) are dissolved in toluene (AR) and diluted to 1 liter. Cap vial immediately, and label the cap. A perfectly dry filter will become completely transparent in the cocktail. Wipe fingerprints from vial before placing it in counter. Count a standard (to determine efficiency) and samples according to instruction of manufacturer. Rarely, if algal concentrations are great or if sample contained large amounts of silt or debris, quenching may occur. Correct for this by preparing a quench curve made up from constant amounts of a ^{14}C standard solution, such as toluene-^{14}C, which is added to vials containing filters through which progressively greater amounts of the quench material–containing water have been filtered. Instructions for quench correction vary with different instruments and must be made according to the manufacturer's instructions.

If you choose the latter method, place the moist filter in the vial as above, and add 20 ml of the following cocktail: dissolve 4 g PPO, 0.1 g dimethyl POPOP, 80 g crystallized naphthalene, and 50 ml *p*-dioxane in toluene (AR), and dilute to 1 liter. Cap the vial, label, and count as above. A disadvantage of this method is the slow solution of the filter in dioxane, which thus cancels out the advantage of the vertical filter.

Calculations

Rates of phytoplankton photosynthesis may be calculated according to the following:

$$P = P_l - P_d$$

and

$$P_l \text{ or } P_d = \frac{r}{R} \times C \times f$$

where

P = phytoplankton photosynthesis in mg carbon per cubic meter
P_l or P_d = carbon uptake per cubic meter in light and dark bottle, respectively

TABLE 8. Factors for conversion of total alkalinity to milligrams of carbon per liter

pH	Temperature (° C)					
	0	5	10	15	20	25
6.0	1.15	1.03	0.93	0.87	0.82	0.78
6.1	0.96	0.87	0.77	0.73	0.70	0.67
6.2	0.82	0.74	0.68	0.64	0.60	0.58
6.3	0.69	0.64	0.59	0.56	0.53	0.51
6.4	0.60	0.56	0.52	0.49	0.47	0.45
6.5	0.53	0.49	0.46	0.44	0.42	0.41
6.6	0.47	0.44	0.41	0.40	0.38	0.37
6.7	0.42	0.40	0.38	0.37	0.35	0.35
6.8	0.38	0.37	0.35	0.34	0.33	0.32
6.9	0.35	0.34	0.33	0.32	0.31	0.31
7.0	0.33	0.32	0.31	0.30	0.30	0.29
7.1	0.31	0.30	0.29	0.29	0.29	0.28
7.2	0.30	0.29	0.28	0.28	0.28	0.27
7.3	0.29	0.28	0.27	0.27	0.27	0.27
7.4	0.28	0.27	0.27	0.26	0.26	0.26
7.5	0.27	0.26	0.26	0.26	0.26	0.26
7.6	0.27	0.26	0.26	0.25	0.25	0.25
7.7	0.26	0.26	0.25	0.25	0.25	0.25
7.8	0.25	0.25	0.25	0.25	0.25	0.25
7.9	0.25	0.25	0.25	0.25	0.25	0.25
8.0	0.25	0.25	0.25	0.25	0.24	0.24
8.1	0.25	0.25	0.24	0.24	0.24	0.24
8.2	0.24	0.24	0.24	0.24	0.24	0.24
8.3	0.24	0.24	0.24	0.24	0.24	0.24
8.4	0.24	0.24	0.24	0.24	0.24	0.24
8.5	0.24	0.24	0.24	0.24	0.24	0.24
8.6	0.24	0.24	0.24	0.24	0.24	0.24
8.7	0.24	0.24	0.24	0.24	0.24	0.24
8.8	0.24	0.24	0.24	0.24	0.23	0.23
8.9	0.24	0.24	0.23	0.23	0.23	0.23

From Saunders, Trama, and Bachmann, 1962.

r = uptake of radioactive carbon in counts per minute = counts

per minute for filtered volume $\times \dfrac{\text{volume of bottle}}{\text{volume filtered}}$

R = ^{14}C inoculated in counts per minute = $2.22 \times 10^6 \times$ micro-curies added $\times$ efficiency of counter

C = inorganic carbon (^{12}C) available in mg per cubic meter = total alkalinity $\times$ conversion factor from Table 8 $\times$ 1,000

f = correction of slower uptake of ^{14}C as compared with ^{12}C = 1.06

For example, assume that 125-ml light bottle is inoculated with 2 microcuries ^{14}C. Fifty milliliters are filtered and give a count of 4,500 per minute at an efficiency of 85%. The pH is 7.5, and the total alkalinity is 100 mg/liter at 20° C.

$$P_l = \frac{4{,}500 \times \dfrac{125}{50}}{2.22 \times 10^6 \,(2)\,(0.85)} \times \begin{array}{l} (100 \times 0.26 \times 1{,}000) \times 1.06 = 82.90 \text{ mg} \\ \text{of carbon uptake in the light per cubic} \\ \text{meter per incubation time} \end{array}$$

The calculation for the corresponding dark bottle uptake proceeds the same. The difference between P_l and P_d approximates net phytoplankton photosynthesis.

COMMUNITY METABOLISM

The metabolic activity of any system consists of two kinds of energy transformation: anabolic (synthetic) and catabolic (respiratory). These are dynamic processes. The metabolic process is both a result and a cause of change. Odum (1956) developed the concepts of autotrophy and heterotrophy in communities. The former term applies to communities in which the long-term synthesis exceeds respiration; the latter term applies to communities in which the long-term respiration exceeds synthesis. These internal processes of synthesis and respiration may be supplemented, to varying degrees, by gain or loss from external sources.

The metabolism of the community is, in reality, the sum of the metabolic activity of each organism in that community. A metabolic study without reference to the individual constituents is relatively easily accomplished and provides much basic information about the state of the community. Carbon dioxide is taken from the water and oxygen added by the process of photosynthesis. Respiration continues throughout the light and dark periods, adding carbon dioxide and utilizing oxygen. As a result, a daily pattern in the concentration of either dissolved gas is observed. A change in the concentration of either gas is used to estimate the rate of formation or the respiration of organic matter.

The diel oxygen method is actually a derivative of the light- and dark-bottle method of Gaarder and Gran (1927). In community studies the entire lake serves as the bottle, with the daylight period analogous to the light bottle and the night hours analogous to the dark bottle. A simplified method (McConnell, 1962) is suitable for small bodies of water with minimum surface agitation.

Apparatus

As required for dissolved oxygen determination

Procedure

Sampling depths and sites are chosen as described under phytoplankton photosynthesis. The procedure involves three samples: two taken on successive evenings and one at the interim morning. These correspond to the high and low concentrations of oxygen found in the lake water. Oxygen concentrations rise during the day but reach a rather steady level in the late afternoon. The afternoon sampling should occur during this plateau. A sample taken 1 hour before sunset best satisfies this requirement. Oxygen concentrations decline during the night. However, upward changes in concentration occur rapidly as soon as light becomes available. Therefore, the time of the dawn sampling is more critical. This time is usually within 1 hour before sunrise.

Oxygen concentrations of the water are determined for

the selected sites and depths at each of these three periods.

Calculations

The three values obtained for each depth at each station are plotted as milligrams per liter against time. A solid line is drawn from the evening value to the following sunrise value and then extrapolated as a broken line for the 24-

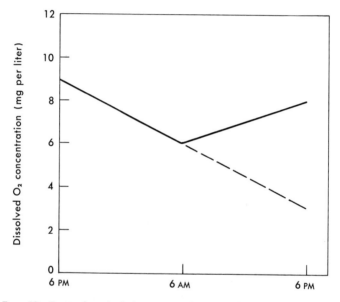

FIG. 23. Example of diel oxygen data used for estimation of community metabolism. At 6 PM of the first day the dissolved oxygen concentration at one station and depth was 9.0 mg/liter. At 6 AM of the following day the dissolved oxygen concentration had declined to 6.0 mg/liter. At 6 PM of the second day, the oxygen concentration increased to 8 mg/liter. The broken line is an extrapolation of the night respiration slope. Total community respiration = 9 mg/liter − 3 mg/liter = 6 mg/liter. Gross community photosynthesis = 8 mg/liter − 3 mg/liter = 5 mg/liter. The ratio of photosynthesis to respiration (P/R) = 0.83.

hour period to represent total decline in oxygen content due to respiration. A solid line is drawn from the dawn value to the second evening value and extended to a point equivalent to 24 hours after the first value. The difference at the 24-hour point between the top and bottom line is gross community photosynthesis as milligrams per liter of oxygen. The difference between the first and the succeeding 24-hour value of the respiration slope is total community respiration as milligrams per liter of oxygen. An example is shown in Fig. 23.

Data as milligrams per cubic meter from each depth may be plotted against depth (in meters) on linear graph paper and integrated by planimetry or by counting squares to calculate the rate of gross photosynthesis or respiration beneath a square meter of lake surface for the depth sampled.

THE
BENTHOS

The benthos are the organisms that inhabit the bottom substrate of lakes, ponds, and streams. These organisms may be artificially divided into two major groups—macrobenthos and microbenthos. Macrobenthos consist of organisms retained by a no. 30 U. S. Series sieve. These procedures are concerned only with macrobenthos that inhabit the mud, sand, gravel, or litter substrate of the lake or stream bottom. Procedures for organisms found on solid bodies projecting above the bottom, such as logs, growing vegetation, and large rocks (aufwuchs), are given later.

Benthic organisms play several important roles in the aquatic community: they are involved in the mineralization and recycling of organic matter produced in the open water above or brought in from external sources, and they are important second and third links in the trophic sequence of aquatic communities. Many benthic insect larval forms are major food sources for small fishes.

Quantitative benthic investigations usually determine the number and kinds of organisms present for the estimation of standing crop and production and serve as indicators of water quality. Benthic organisms are especially useful in pollution studies, and indices of their diversity are widely used for estimation of the degree of pollution. The procedures given in this chapter include methods for the collection, concentration, separation, identification, and enumeration of these organisms.

COLLECTION OF BENTHIC SAMPLES
Lentic

A variety of lentic bottom dredges (grabs) are available. Each was originally designed for one specific sampling problem. Several common collecting instruments are the Ekman dredge, the Petersen dredge, and the Ponar dredge (Fig. 24). The Ekman dredge is the easiest to use in that it is light and relatively easy to "set." However, its use is limited to soft mud, silt, or finely divided sand bottoms. Because of its relatively small sampling area, one must take many replicate samples. For sampling where the bottom material is compacted or consists of gravel, rock, or organic litter substrate, the Petersen or the Ponar dredge is preferred. These dredges are quite heavy and therefore normally require a winch. The weight will vary between 35 and 70 lb., depending on the number of weights applied. Small obstacles to the closing of these dredges are crushed by the jaws, whereas the same materials would block the operation of the Ekman dredge.

A precaution to be noted with any of these types of dredges is the possible disturbance of the surface film of the bottom substrate as the dredge is lowered. Although the Ekman dredge employs a hinged top that allows the column of water to pass through, there is always some displacement of water. With the Petersen dredge, this displacement is considerable. It produces an outward blastlike current on the finely divided substrate just before the dredge settles. To overcome this problem the dredge must be lowered gently through the last ½ meter.

Although the closing of the Petersen or Ponar dredge takes a considerably deeper bite than does the Ekman, each samples only the first few centimeters of substrate. It has been demonstrated that in many waters the bottom fauna is distributed for depths considerably below the bite of these dredges. In shallow waters (up to 3 meters), it is possible to fit an Ekman dredge with a solid rod mechanism rather than the rope and messenger used in deeper waters. With this modification, one can lower the Ekman dredge onto the bottom and then force it deeper into the sediments as it is being tripped. However, in any investigation it is probable that a certain portion of the deep-burrowing benthos is missed by any of these commonly used sampling techniques.

When these dredges are raised, it is important that they not be lifted clear of the water and allowed to drain, as small organisms will be lost through the small seams from which the water runs. Therefore, the best procedures require a lifting of the dredge from the surface of the water while slipping a bucket beneath the dredge to catch these materials.

FIG. 24, A. Commonly used benthos dredges in the open position.

FIG. 24, B. Commonly used benthos dredges in the closed position. The Ponar (far left in **B**) and the Petersen (second from left in **B**) dredges are used when the lake bottom is compacted or stony or contains woody debris. The screen material on top of the Ponar dredge allows water to pass as the dredge is lowered, thus lessening the wave shock disturbance of light-bodied benthos at the mud-water interface. The Ponar dredge may occasionally become jammed by small gravel lodged between the end plate and the closing jaw. The Petersen dredge as shown has two additional weights bolted to each jaw to produce a deeper bite. One or both may be removed for softer sediments. Both dredges at the right of **B** are Ekman dredges. The one to the far right has been modified by welding a ¾-inch pipe nipple to the trip mechanism. This permits the dredge to be attached to a 10- to 12-foot segment of pipe for use in shallow water. Used this way, the sampler may be thrust deeper into the sediments to provide a better sample than that taken by the standard Ekman dredge (second from right). The standard Ekman dredge is operated by line, and the spring-loaded jaws are tripped by a brass messenger to snap closed. Both Ekman dredges have hinged top plates that open on descent, allowing water to pass and minimizing wave shock disturbance.

Apparatus

Dredge, line, and messenger
Two or more metal 10-quart pails
Metal laundry tubs
Large-diameter no. 30 screen
Wide-mouthed bottle (1 pint to ½ gallon)

Reagents

Supply of neutral formalin

Procedure

1. Select sampling sites representative of the area to be sampled, and determine number of replications necessary. Be sure to record number of replicate samples taken.

2. Set dredge trip mechanism after securing other end of line to some fixture of the boat. These dredges are especially heavy when loaded, and precautions should be taken to avoid dropping and loss.

3. Lower dredge to sampling site, lowering slowly through the final ½ meter. If depth is unknown, lower to bottom and then raise, move 1 or 2 meters laterally, and relower gently. Trip dredge by dropping messenger for the Ekman or by allowing the line to slacken for the Petersen or Ponar.

4. Lift the filled dredge to the surface with a smooth, even motion to avoid jarring out contents, and swing into a pail as mentioned above.

5. Dump contents of dredge and pail into a washtub. If the boat is large, the following procedures may be done in the boat; if not, return filled washtub to shore and do following procedures there. Repeat the sampling procedure until all of the replicates have been taken.

6. Reach into the tub, and break up all compacted particles. These are especially common in areas of clay soils. Continue working the materials in the tub until it is of a fine homogeneous consistency.

7. Pour the contents of the tub (use a pail to dip when the tub is full) into the no. 30 screen, holding the screen

Text continued on p. 130.

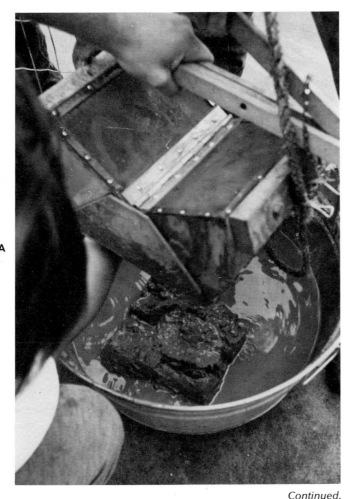

A

Continued.

Fig. 25. A, Contents of a Ponar sample are emptied into a large tub. **B,** Water is added, and large particles are broken up by hand. **C,** This slurry is then poured through a benthos screen. **D,** Sieving of the screen contents is accomplished by sloshing in and out of the water with a twisting or swirling motion. **E,** Mud and sand will pass through the screen, leaving a mixture of debris and benthic organisms on the screen. This mixture is usually preserved and returned to a laboratory for further separation and benthos analysis.

B

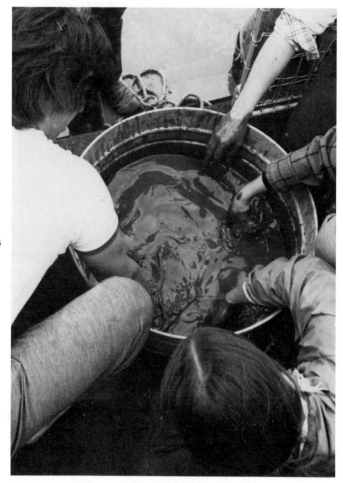

FIG. 25, cont'd. B, Adding water.

C

Continued.

Fig. 25, cont'd. C, Pouring slurry through benthos screen.

D

FIG. 25, cont'd. D, Sieving screen contents.

FIG. 25, cont'd. E, Debris and benthic organisms on screen.

over the edge of the boat. Continue this step until all the contents of the tub have been poured through the screen and the tub and any pails used have been rinsed several times with lake water.

8. The materials remaining on the screen are your collected samples. Use a sloshing, twisting, and swirling motion as you thrust the screen up and down into the water, but do not let the water run over the top of the screen. This will wash smaller materials through the sieve and at the same time will tend to concentrate the materials at one edge if you work holding the screen on a slant.

9. When all materials are concentrated in one area, scrape these into a wide-mouthed bottle, wash screen again, and place any additional materials into the bottle, picking out the last few remains with fine forceps if necessary.

10. Preserve this material with adequate neutral formalin to produce a 5% to 10% solution (one part of formalin to nine parts water, ooze, and organisms).

Lotic

Sampling from running waters differs in both the nature of the organisms collected and the means used for the collection. The distinction between benthos and aufwuchs is poorly defined in running waters. Because of the scouring action of the current, soft sediments are rarely found. Organisms of running waters are either heavy bodied or have special means of attachment. Finally, because of the lack of fine sediments, the distribution of organisms with depth in the sediment is usually greater in running waters than in standing waters.

The most commonly used sampling device for running waters is the Surber stream bottom sampler (Fig. 26). This device has a major drawback in that its use is restricted to waters of less than 1 foot in depth and of slow to moderate velocity. Because quantitative sampling devices are lacking for other types of flowing water, the only method suitable for such situations is an indirect measure-

FIG. 26. The Surber stream sampler is used for quantitative benthos analysis of shallow streams and rivers. Placed parallel with the current, the bottom substrate is stirred within the area of the frame. Dislodged organisms are washed into the net. (Courtesy Wildlife Supply Company, Saginaw, Michigan 48602.)

ment of the benthic fauna by measurement of drift using a series of drift nets.

Apparatus

Surber stream bottom sampler
Garden trowel
Stiff bristle brush (fingernail type)
Wide-mouthed bottle (1 pint to ½ gallon)

Reagents

Neutral formalin

Procedure

1. Select sampling site representative of area desired and with no depth greater than height of sampling frame.

The velocity of flow must not be so great as to cause a "pressure head" of water to flow around mouth of net.

2. Wade into water from downstream, and place net with the mouth facing upstream in an undisturbed area. Be sure there is no disturbance of the bottom substrate upstream from the net, or organisms will be dislodged and washed into the net.

3. Lower square foot frame on substrate, and hold it in place. Pick up larger rocks or bits of substrate, and while holding them in mouth of net, brush them free of all organisms, allowing current to carry them into net. Discard these rocks outside the frame.

4. When all larger bits have been brushed, use garden trowel to stir up all the substrate within the square foot frame. Be careful that the stirring motion is toward center of frame so that organisms that are dislodged will be carried into mouth of net and not around it. Attempt to stir this area thoroughly and to a uniform depth.

5. Depending on the number of replicates desired, move the sampler to other sites, and repeat the above procedure.

6. After all replicates have been taken, empty contents of net into a wide-mouthed jar. The net must be turned "wrong side out" and organisms that have attached to the fabric picked from net and placed into jar.

7. Add sufficient water to cover the substrate and organisms in the bottle and then sufficient formalin to make a 5% to 10% solution. Label jar, indicating number of replicates, and return to the laboratory.

LABORATORY PROCESSING OF BENTHIC SAMPLES

The collected samples contain a mixture of mud, rocks, sand, debris, and the desired organisms. The next step is to remove the organisms from the unwanted material and separate them into similar taxonomic groupings for final identification and enumeration. A variety of laboratory techniques, each specifically suited for a particular type of sample, is used by different investigators. Techniques described below are general and should be modified where such a modification would lend improvement. It is important for the student to keep in mind that the procedures used must be nonselective so that the final list will be truly representative of the habitat sampled.

Apparatus

Graded series of soil sieves
Shallow white enamel pans
Fine-tipped, gentle action forceps
Large glass funnel (8 to 10 inches in diameter) with large drain opening (1 to 2 inches) fitted with flexible rubber tubing and hose clamps
Dissecting microscope with good illumination
Series of small vials or baby food jars

Reagents

Table sugar (several pounds)
4% neutral formalin solution

Procedure

If the collected sample contains much extraneous material, such as gravel or organic debris (which is usually the case with Surber samples), proceed as follows. If the sample has been well screened in the field, omit the laboratory screening procedures and go directly to step 3.

1. Place the graded series of soil sieves into a sink, and empty sample or part of sample into top screen. Allow a gentle stream of water to wash through screen to thoroughly wash and sort materials into different screens. Be careful that bottom screens do not plug, causing a back-flushing and loss of organisms over the top.

2. Remove upper screen of very coarse materials, and pick through it for very large organisms. As organisms are separated, place them into vials or baby food jars containing 4% formalin according to their approximate taxonomic groupings.

3. Organisms from screens may be separated from the

rubble by several methods. Each is best suited for particular types and frequency of organisms and the nature of the rubble. Three methods are given below.

3a. Empty each screen with its contained rubble and organisms into a shallow enamel pan, add enough water to disperse the materials, and using forceps, pipette, or small metal scooping screen, take off organisms from rubble, stirring frequently to ensure that all are loosened and separated from the debris. Check material from finer screens under a dissecting microscope to be sure that no small organisms are being missed.

3b. Place materials from screens in white enamel pan, and flood with sugar solution. This solution should be made up of table sugar and tap water and should have a specific gravity of 1.12 (approximately 300 gm of sugar per liter). Make this deep enough to cover the debris by at least 1 inch. Stir debris, allow currents to stop, and soft-bodied forms will float to the surface, where they may be scooped off. Repeated stirrings are necessary to allow materials to separate. Complete this process within 15 minutes, or organisms will begin to sink. Heavy-bodied organisms, such as fingernail clams, stone-cased caddis flies, and oligochaetes, must be picked off the bottom. Pour material remaining in tray back through a screen of equal or smaller size, and collect the sugar solution, which should be saved and used many times. Flood debris with tap water, and allow it to soak for 15 minutes; then repeat sugar flotation procedure after tap water has been poured off. This second floating serves as a check on your separation. Always make the final inspection with a dissecting microscope. When sugar solution becomes strongly discolored so that it is not possible to see through it to bottom of pan, it may be clarified by running through a pad of gauze or coarse filter paper. This sugar solution may be reused (until it is no longer possible to clear it) by making only necessary adjustments to the specific gravity by addition of more sugar.

3c. The following procedure is also a sugar separation

FIG. 27. Sugar flotation may aid in separation of benthos from heavy debris. Here a sample is added to a funnel containing a sugar solution. This solution is stirred well and then allowed to settle. Debris and an occasional heavy-bodied organism are drained into one container, while the supernatant fluid with floating organisms are separated into a second container. This separation is repeated on the first settlings.

technique. Mount large funnel in ring stand. The rubber tubing is connected to the bottom and closed with pinch clamp. Fill this funnel approximately half full of the 1.12–specific gravity sugar solution as in 3b. Empty contents of one screen into this funnel, and stir. Allow contents to settle, and most debris will sink into neck of funnel. Squeeze the soft rubber tubing in a pumping motion to set up currents, which will tend to dislodge organisms trapped in the debris. Allow contents to separate for a few seconds, then skim off organisms from top of funnel. Drain funnel through tubing into two vessels in the same manner one would use a separatory funnel, placing heavy debris in one and supernatant into a second. Be sure to wash down sides of funnel with sugar solution into the second container. The supernatant can then be passed through a fine screen to collect any small organisms that were not skimmed from the top. After this has been done, repeat procedure on same material that has been collected in first container.

4. In every case, after organisms have been separated from the debris, they should be washed free of any sugar solution, placed in approximate taxonomic groupings (if possible), and preserved in 4% neutral formalin.

5. Identification of most bottom fauna requires the use of a dissecting microscope with good illumination. Pennak's *Fresh-water Invertebrates of the United States* is a good starting point for identification. Edmondson's *Fresh-water Biology* should also be available.

6. After organisms have been identified, count them by taxonomic group, and report as numbers per area. Determine the area sampled by the dredge used, and multiply this by the number of replicates taken to get the total area sampled. Data should be standardized and reported as number of organisms per square meter.

7. Further processing may be desirable in that some data is better expressed as biomass. In this case it is necessary to wash the organisms in tap water to remove the

formalin and blot dry with absorbent tissue for a wet "live" weight, or place in shallow dishes and dry in a laboratory oven at 103° to 105° C, and cool in a desiccator before weighing for dry weight.

AUFWUCHS

The aufwuchs may be considered a special case of the benthos. The aufwuchs make up any community that develops upon the surface of any submerged substrate. Many investigators use this term interchangeably with the term periphyton. The aufwuchs community developing upon hard or permanent, usually rock substrates can be distinguished as epilithic aufwuchs; the community developing upon organic, usually plant substrates are epiphytic.

By definition, then, this community will have a highly variable composition; and although it may include such large members as snails and insect larvae, investigators are more typically concerned with the microscopic members. Protozoa, rotifers, and nematodes are especially common animals, whereas the diatoms and blue-green algae are the common plants. Colonial, stalked, and gelatinous forms are common among both plants and animals. Many species in the typical aufwuchs community are delicate, soft-bodied organisms. Therefore, special care must be taken in handling and preserving them. In fact, it is often preferable to examine this community alive and undisturbed.

Although relatively difficult to investigate, this community provides an excellent biological detector of water quality. Because of their sessile nature, the aufwuchs cannot move to escape onslaughts of pollution; thus these organisms are constant, round-the-clock detectors of pollution.

Study of the aufwuchs may be undertaken from natural substrates or by colonization of artificial substrates. Whereas artificial substrates have many practical advantages, especially for quantitative studies, it is probable that a "natural" community never develops on an artificial substrate in any reasonable period of time.

Once collected, identification and counting of aufwuchs proceeds much the same as for plankton. The Sedgwick-Rafter cell may be used on well-mixed samples for larger forms; sedimentation techniques and gentle membrane filtration may be used for smaller forms.

Collection methods—natural substrates

Qualitative sampling may be done by gently scraping the material from any submerged surface into a small collection jar or vial. If examination cannot be accomplished within 1 day, preservation in 5% formalin is necessary. Quantification of natural substrates is difficult due to their irregular surface area. Obviously, the surface area of a regular object, such as the cylindrical stem of a bullrush, may be quantified if the entire surface area is scraped into the collecting jar. Other, more irregular surfaces may be quantified as follows.

Apparatus

Isolating cylinder (A large-diameter cork borer is ideal; however, any type of cylinder that can be beveled to a sharp edge is satisfactory.)
Stiff fingernail brush
Scraping device, such as spatula or knife
Vials or small baby food jars

Reagents

Supply of 5% formalin

Procedure

1. Select the natural substrate—wood, stone, and so forth—that appears to have been in its present position for some time. Remove from water, but keep surface wet.

2. Using the cork borer or other isolating cylinder, press down a "randomly" selected portion of the substrate, and while holding borer firmly in place, brush away

all aufwuchs around the cylinder for an area of at least 2 inches. Wash this area off with water to be certain that all surrounding organisms have been removed. At this point, you have isolated an island of aufwuchs of known area.

3. Gently scrape this isolated island into the small collection vial with a small amount of water. If possible, return vial to the laboratory while keeping it cool, and proceed with identification and counting of the living material. If identification and counting must be postponed, preserve organisms in 5% formalin.

Collection methods—artificial substrates

Many types of artificial substrates, including concrete blocks, bricks, and building tiles, have been used for aufwuchs studies. These materials have the advantage of having regular, known area surfaces that can be entirely scraped clean for quantitative determinations. Another type of artificial substrate that allows the study of the aufwuchs community without damage by scraping is the glass microscope slide.

Apparatus

Support for microscope slides. Any type of device that will rigidly hold a series of separated microscope slides in a vertical position, and if in a flowing system, with the surfaces parallel to the current's flow, will suffice. This is accomplished by using a wooden microscope slide file box from which both the top and bottom are removed, leaving only the ends and sides with their cut divisions for separating the slides. The top and bottom are covered with rectangular pieces of ¼- to ½-inch mesh hardware cloth. Once the slides have been inserted in the box, the hardware cloth is held in position by heavy rubber bands or cord.

Large-mouthed, 2-quart to 1-gallon jars of sufficient size to contain the slide box rack.

Ropes, anchors, and floats necessary to suspend slide box artifical substrate rack.

Reagents

Supply of formalin sufficient to produce a 4% to 5% solution when added to the large-mouthed container full of water.

Procedure

1. Number microscope slides, using a diamond-tipped scribe, permanent ink, or paint. Place slides in rack, and firmly bind so that they will not be dislodged through a long colonization period.

2. Suspend artificial substrate rack in the lake or river in such a way that slides are held in a vertical position, and if there is current, with flat surface parallel to direction of current. A series of these slides through various depths may be desired, since distinct depth distributional patterns occur among the aufwuchs, with autotrophic components dominating near the surface and heterotrophic components dominating in the deeper waters.

3. Allow artificial substrates to colonize for a minimum of 2 weeks. Colonization time may need to be longer, depending on the season of the year and water temperature, as well as the nature of the study.

4. After colonization, gently remove entire sampler from water, and place in large-mouthed jar to be returned to the laboratory for live examination if possible, or preserve in 4% formalin if necessary. A preserved slide has the advantage in that one may work directly with oil immersion on the preserved slide without a cover slip.

5. In the laboratory scrape one surface of microscope slide clean with a razor blade. This surface will be the "bottom" and will be placed on the microscope stage. While keeping upper surface wet, scan entire microscope slide with both a dissecting microscope and a compound microscope.

6. Take quantitative counts, using a Whipple eyepiece

disk, counting all organisms contained in random fields (by using a mechanical stage, you may take strip counts). It is also possible to scrape off all the aufwuchs components on the surface of the slide except for a known area and then take a total count of the larger organisms for the known area.

7. Biomass may be determined by scraping entire surface of microscope slide with a razor blade and collecting the aufwuchs in an evaporating dish, where the organisms are dried and weighed, using the techniques described under Residue (Chapter 2).

DIVERSITY OF BENTHOS

Diversity indices relate the number of kinds of organisms to the total number of organisms and in some cases to the number of individuals of each kind. Such indices may be applied to any biotic community but have had widest application with the benthos. A statement of species diversity has become widely used as an indicator of water quality—low diversity indicates water of low quality. Undisturbed natural communities are assumed to have a high diversity; that is, a relatively large number of species, with no species having disproportionately large numbers of individuals.

A number of calculations, each with specific applications, have been developed to express diversity. One of the simplest is: diversity $(D) = m/N^{\frac{1}{2}}$, where m is the total number of species and N is the total number of individuals. However, any of these methods requires a taxonomic knowledge of the organisms. One type of diversity index, the sequential comparison index, does not have this requirement, thus permitting the most inexperienced to use it (Cairns and others, 1968).

Sequential comparison index
Procedure

1. Combine all organisms from a given collection. Use some method to arrange the organisms randomly in a

FIG. 28. A portion of a randomly arranged row of stream benthos for calculation of the sequential comparison index. This segment of 11 organisms contains 8 obviously different runs. A difficult judgment must be made to determine if the small mayfly (fifth from left) is different enough from the preceding larger mayfly to count it as a separate run.

straight line. For example, you might add a small amount of water and empty the organisms into a rectangular white enamel pan. Wash down to one side with a stream of water from a squeeze bottle, then blot off excess water.

2. Begin at one end of the row of organisms. Pick them out one at a time and decide if this organism is different from the preceding organism. If it is not, go on to the next one; if it is different, record this as the start of a new "run." Therefore, a new run occurs when any organism is different from the preceding one. At the end of the row, count up the number of runs and the total number of organisms.

3. Calculate the diversity index (DI):

$$DI = \frac{\text{number of runs}}{\text{number of organisms}}$$

4. If the sample size is excessively large (greater than 250 organisms), stop at the end of the first 50 organisms

and calculate DI; that is, number of runs/50. Then continue for the next 50 organisms and recalculate: number of runs/100. On graph paper make a plot of calculated DIs on the ordinate against number of specimens on the abscissa. Continue calculating and plotting at the end of each 50 organisms until the curve obtained becomes obviously asymptotic. At that point there is no advantage in continuing. A statistical analysis of such data indicates that when sufficient organisms are present in the sample as described in this step, a rerandomization and repeat of the procedure is sufficient to produce a $\overline{DI}$ with a 95% confidence interval equal to ±10% of calculated $\overline{DI}$.

PRODUCTION OF BENTHOS

Methods for the measurement of secondary production of benthos are not as generally applicable as those for primary production. Methods for measurement of secondary production must be quite specific for each group of organisms. Differences in length of life span, presence or absence of juvenile instars, carried or demersal eggs, size (age) specific or nonspecific mortality, population size, and turnover rates all must be considered. Consequently, it is difficult to present a single method. The following is a method of applicability to organisms with a single generation per year (univoltine), for which the turnover rate, birth rate, or death rate is unknown. It may be applied to a single taxon or to the collective benthic community.

Procedure

1. Collect organisms quantitatively from a known area, and separate as described above. Prepare for dry weight determinations.

2. Weigh dried organisms, and tabulate numbers into weight groups. Considering smallest and largest organisms, determine magnitude of weight groupings so that organisms can be separated into 4 to 10 groups (see Table 9).

3. Repeat this procedure for a series of equally spaced dates for the same habitat. If calculation of animal production is desired, this series might be twelve monthly samplings.

4. Tabulate as in the example shown in Table 9 for only four size groups and five collections.

TABLE 9. Example of tabulation for calculation of benthic production

Size group (mg)	Collections (nos./area)					N	$\bar{n}$
	1	2	3	4	5		
0.0–0.49	150	130	100	70	20	470	94
0.5–0.99	60	40	25	5	2	132	26
1.0–1.49	50	40	20	10	2	122	24
1.5–1.99	10	10	9	3	0	32	6

Calculations

Calculate production according to:

$$P_B = i \sum (\bar{n}_j - \bar{n}_{j-1}) \frac{\overline{w}_j - \overline{w}_{j-1}}{2}$$

where

P_B = biomass production per m² per time
i = number of collections
n_j = number of organisms of size group j per m²
$\bar{n}_j$ = n_j/number of collections
$\overline{w}_j$ = mean weight of size group j

therefore

$$P_B = [5 (94 - 26) \left(\frac{0.25 + 0.75}{2} \right) + (26 - 24)$$
$$\left(\frac{0.75 + 1.25}{2} \right) + (24 - 6) \left(\frac{1.25 + 1.75}{2} \right)$$
$$+ 6(2.0)] = 375 \text{ mg per m}^2 \text{ per 5 time units}$$

USEFUL CONVERSION FACTORS

To convert	Into	Multiply by
acre	hectare	0.4047
acres	square feet	43,560.0
acres	square meters	4,047.0
acre-feet	cubic feet	43,560.0
acre-feet	gallons	3.259×10^5
centigrade	fahrenheit	$(°C \times 9/5) + 32$
centimeters	feet	3.281×10^{-2}
centimeters	inches	0.3937
cubic centimeters	cubic feet	3.531×10^{-5}
cubic centimeters	cubic inches	0.06102
cubic centimeters	cubic meters	10^{-4}
cubic centimeters	gallons (U. S. liquid)	2.642×10^{-4}
cubic feet	cubic meters	0.02832
cubic feet	gallons (U. S. liquid)	7.48052
cubic feet	liters	28.32
cubic inches	cubic centimeters	16.39
cubic inches	cubic meters	1.639×10^{-5}
cubic inches	liters	0.01639
cubic meters	cubic feet	35.31
cubic meters	cubic yards	1.308
cubic meters	gallons (U. S. liquid)	264.2
ergs	gram calories	0.2389×10^{-7}

To convert	Into	Multiply by
ergs	kilocalories	2.389×10^{-11}
ergs/second	kilocalories/minute	1.433×10^{-9}
fathom	meter	1.828804
fathoms	feet	6.0
feet	centimeters	30.48
feet	kilometers	3.048×10^{-4}
feet	meters	0.3048
foot-candle	lumen/square meter	10.764
gallons	cubic centimeters	3,785.0
gallons	cubic feet	0.1337
gallons	cubic meters	3.785×10^{-3}
gallons	liters	3.785
gallons of water	pounds of water	8.3453
gallons/minute	cubic feet/second	2.228×10^{-3}
gallons/minute	liters/second	0.06308
grams	grains	15.43
grams	ounces (avoirdupois)	0.03527
grams	ounces (troy)	0.03215
grams	pounds	2.205×10^{-3}
grams/centimeter	pounds/inch	5.600×10^{-3}
grams/liter	parts/million	1,000.0
grams/square centimeter	pounds/square foot	2.0481
gram calories	ergs	4.1868×10^{7}
hectares	acres	2.471
hectares	square feet	1.076×10^{5}
inches	centimeters	2.540
inches	meters	2.540×10^{2}
joules	ergs	10^{7}
joules	kilocalories	2.389×10^{4}
kilograms	pounds	2.205
kilograms/cubic meter	pounds/cubic foot	0.06243
kilograms/meter	pounds/foot	0.6720
kilograms/square meter	pounds/square foot	0.2048
kilometers	centimeters	10^{5}
kilometers	feet	3,281.
kilometers	miles	0.6214

To convert	Into	Multiply by
kilometers/hour	feet/second	0.9113
knots	statute miles/hour	1.151
liters	cubic feet	0.03531
liters	gallons (U. S. liquid)	0.2642
liters	quarts (U. S. liquid)	1.057
liters/minute	cubic feet/second	5.886×10^4
lumens/square foot	foot-candles	1.0
lux	foot-candles	0.0929
meters	feet	3.281
meters	inches	39.37
meters	miles (statute)	6.214×10^4
meters	yards	1.094
meters/minute	feet/second	0.05468
microns	meters	1×10^6
miles (statute)	kilometers	1.609
miles (statute)	meters	1,609.
miles (statute)	miles (nautical)	0.8684
milligrams/liter	parts/million	1.0
millimeters	feet	3.281×10^{-3}
millimeters	inches	0.03937
millimicrons	meters	1×10^{-9}
million gallons/day	cubic feet/second	1.54723
ounces	grams	28.349527
ounces	pounds	0.02957
parts/million	grains/U. S. gallon	0.0584
pounds	grains	7,000.
pounds	grams	453.5924
pounds	kilograms	0.4536
pounds	ounces	16.0
pounds of water	cubic feet	0.01602
pounds of water	cubic inches	27.68
pounds of water	gallons	0.1198
pounds/foot	kilograms/meter	1.488
pounds/inch	grams/centimeter	178.6
pounds/square foot	inches of mercury	0.01414
quarts (liquid)	cubic centimeters	946.4
quarts (liquid)	cubic feet	0.03342

To convert	Into	Multiply by
quarts (liquid)	cubic meters	9.464×10^{-4}
quarts (liquid)	liters	0.9463
square centimeters	square feet	1.076×10^{-3}
square centimeters	square inches	0.1550
square centimeters	square meters	0.0001
square feet	acres	2.296×10^{-3}
square feet	square centimeters	929.0
square feet	square meters	0.09290
square inches	square centimeters	6.452
square kilometers	acres	247.1
square kilometers	square feet	10.76×10^{6}
square kilometers	square miles	0.3861
square meters	acres	2.471×10^{-4}
square meters	square centimeters	10^{4}
square meters	square feet	10.76
square meters	square miles	3.861×10^{-7}
square meters	square yards	1.196
square miles	square kilometers	2.590
square miles	square meters	2.590×10^{6}
square yards	square meters	0.8361
watts	ergs/second	107.
watts	kilocalories/minute	0.01433
watt-hours	ergs	3.60×10^{10}
watt-hours	gram calories	859.85
yards	centimeters	91.44
yards	kilometers	9.144×10^{-4}
yards	meters	0.9144

SOURCES OF LIMNOLOGICAL APPARATUS AND SUPPLIES

This partial list implies no endorsement of product or service. A more complete listing is *Sources of Limnological and Oceanographic Apparatus and Supplies* (special publication no. 1, third revision) available from the American Society of Limnology and Oceanography.

Reagents

Harleco
   60th and Woodland Ave.
   Philadelphia, Pa. 19143

Test kits and reagents

Hach Chemical Co.
   Box 907
   Ames, Iowa 50010
Hellige, Inc.
   877 Stewart Ave.
   Garden City, N. Y. 11533

Radioisotopes

Amersham-Serale Corp.
   2636 S. Clearbrook Dr.
   Arlington Heights, Ill. 60005

New England Nuclear
  575 Albany St.
  Boston, Mass. 02118

General sampling and field equipment

Wildco
  2200 S. Hamilton St.
  Saginaw, Mich. 48602
Kahl Scientific Instrument Co.
  P. O. Box 1166
  El Cajon, Calif. 92022
Hydro-Bios Apparatebau
  Wismarerstrasse 14
  23 Kiel, Germany
Hydro Products
  4930 Naples Place
  San Diego, Calif. 92110
Foerst Mechanical Specialties
  2407 N. St. Louis Ave.
  Chicago, Ill. 60647

Submarine photometers, thermometers, and other supplies

Montedoro-Whitney Corp.
  P. O. Box 1401
  San Luis Obispo, Calif. 93401
Kahl Scientific Instrument Co.
  P. O. Box 1166
  El Cajon, Calif. 92022

Membrane filters and supplies

Millipore Corp.
  Bedford, Mass. 01730

Nets and screens

General Biological Supply (Turtox)
  8200 S. Hoyne Ave.
  Chicago, Ill. 60620

REFERENCES

American Public Health Association. 1971. Standard methods for the examination of water and wastewater. American Public Health Association, Inc. New York. 874 pp.

Brown, E., M. W. Skougstad, and M. J. Fishman. 1970. Methods for collection and analysis of water samples for dissolved minerals and gases. U. S. Geological Survey, U. S. Dept. of the Interior, Washington, D. C. 166 pp.

Cairns, J., D. W. Albaugh, F. Busey, and M. D. Chanay. 1968. The sequential comparison index—a simplified method for nonbiologists to estimate relative differences in stream pollution studies. J. Water Poll. Control Fed. **40**:1607-1613.

Correll, D. S., and H. B. Correll. 1972. Aquatic and wetland plants of southwestern United States. Environmental Protection Agency, Washington, D. C. 1777 pp.

Eddy, S., and H. C. Hodson. 1961 Taxonomic keys to the common animals of north central states. Burgess Publishing Co., Minneapolis. 162 pp.

Edmondson, W. T., ed. 1959. Fresh-water biology, 2nd ed. John Wiley & Sons, Inc., New York. 1248 pp.

Edmondson, W. T., and G. G. Winberg. 1971. A manual on methods for the assessment of secondary productivity in fresh waters. International Biological Program Handbook No. 17. Blackwell Scientific Publications Ltd., Oxford. 358 pp.

Ellis, M. M., B. A. Westfall, and M. B. Ellis. 1948. Determination of water quality. Research Report No. 9. Fish and Wildlife Service, U. S. Dept. of the Interior, Washington, D. C. 122 pp.

Environmental Protection Agency. 1971. Methods for chemical

analysis of water and wastes. Environmental Protection Agency, Water Quality Office, Analytical Quality Control Laboratory, Cincinnati. 312 pp.

Eyles, D. E., and J. L. Robertson. 1963. A guide and key to the aquatic plants of the southeastern United States. Public Health Bulletin No. 86. Bureau of Sport Fisheries and Wildlife, Washington, D. C. 151 pp.

Fassett, N. C. 1957. A manual of aquatic plants. University of Wisconsin Press, Madison. 405 pp.

Gaarder, T., and H. H. Gran. 1927. Investigations of the production of plankton in the Oslo Fjord. Rapports et Procès—Verbaux des Réunions Conseil Permanent International pour l'Exploration de la Mer **42**:1-48.

Golterman, H. L., and R. S. Clymo. 1969. Methods for chemical analysis of fresh waters. International Biological Program Handbook No. 8. Blackwell Scientific Publications Ltd., Oxford, 188 pp.

Hutchinson, G. E. 1957. A treatise on limnology, vol. 1. John Wiley & Sons, Inc., New York. 1015 pp.

Lind, O. T., and R. S. Campbell. 1969. Comments on the use of liquid scintillation for routine determination of ^{14}C activity in production studies. Limnol. and Oceanogr. **14**:287-289.

Maciolek, J. 1962. Limnological organic analysis by quantitative dichromate oxidation. Research Report No. 60. Bureau of Sport Fisheries and Wildlife. 61 pp.

McConnell, W. J. 1962. Productivity relations in carboy microcosms. Limnol. and Oceanogr. **7**:335-343.

Moore, E. W. 1939. Graphic determination of carbon dioxide and the three forms of alkalinity. J. Amer. Water Works Assn. **31**:51.

Murphy, J., and J. Riley. 1962. A modified single solution method for the determination of phosphate in natural waters. Anal. Chim. Acta. **27**:31-36.

Needham, J. G., and P. R. Needham. 1962. A guide to the study of fresh-water biology, 5th ed. Holden-Day, Inc., San Francisco. 108 pp.

Odum, H. T. 1956. Primary production in flowing waters. Limnol. and Oceanogr. **1**:102.

Odum, H. T., W. McConnell, and W. Abbott. 1958. The chlorophyll "A" of communities. Publ. Inst. Mar. Sci., Texas **5**:56-96.

Patrick, R., and C. W. Reimer. 1966. The diatoms of the United

States, vol. 1. Monographs of the Academy of Natural Sciences, Philadelphia. 688 pp.

Pennak, R. W. 1953. Fresh-water invertebrates of the United States. The Ronald Press Co., New York. 769 pp.

Poole, H. H., and W. R. G. Atkins. 1926. On the penetration of light into sea water. J. Mar. Biol. Assn. U. K. **14:**177-198.

Prescott, G. W. 1962. Algae of the western Great Lakes area. William Brown Co., Publishers, Dubuque, Iowa. 977 pp.

Prescott, G. W. 1970. How to know the fresh water algae, 2nd ed. William C. Brown Co., Publishers, Dubuque, Iowa. 348 pp.

Rainwater, F. H., and L. L. Thatcher. 1960. Methods for collection and analysis of water samples. U. S. Geological Water Supply Paper 1454. U. S. Government Printing Office, Washington, D. C. 301 pp.

Ricker, W. E. 1968. Methods for assessment of fish production in fresh waters. International Biological Program Handbook No. 3. Blackwell Scientific Publications Ltd., Oxford. 326 pp.

Ryther, J. H., and C. S. Yentsch. 1957. The estimation of phytoplankton production in the ocean from chlorophyll and light data. Limnol. and Oceanogr. **2:**281-286.

Saunders, G. W., F. B. Trama, and R. W. Bachmann. 1962. Evaluation of a modified C-14 technique for shipboard estimation of photosynthesis in large lakes. Great Lakes Res. Div. Publ. No. 8, University of Michigan, Ann Arbor. 61 pp.

Schwoerbel, J. 1970. Methods of hydrobiology (fresh-water biology). Pergamon Press Ltd., Oxford. 200 pp.

Small, L. F. 1961. An optical density method of measuring phytoplankton standing crop. Iowa State J. Sci. **35:**343.

Smith, G. M. 1950. The fresh water algae of the United States, 2nd ed. McGraw-Hill Book Co., New York. 719 pp.

Stephens, K. 1963. Determination of low phosphate concentrations in lake and marine waters. Limnol. and Oceanogr. **8:**361-362.

Strickland, J. D. H., and T. R. Parsons. 1968. A practical handbook of sea water analysis. Fisheries Research Board of Canada, Ottawa. 311 pp.

Vollenweider, R. A. 1969. A manual on methods for measuring primary production in aquatic environments. International Biological Program Handbook No. 12. Blackwell Scientific Publications Ltd., Oxford. 213 pp.

Welch, P. S. 1948. Limnological methods. McGraw-Hill Book Co., New York. 381 pp.

Wolfe, D. A., and C. L. Schelske. 1967. Liquid scintillation and geiger counting efficiencies for carbon-14 incorporated by marine phytoplankton in productivity measurements. J. Cons. Perm. Inter. Explor. Mer. **31:**31-47.